S0-AKJ-166

# COMPUTER-ASSISTED INSTRUCTION IN CHEMISTRY

*(IN TWO PARTS)*

Part B: Applications

COMPUTERS IN
CHEMISTRY AND INSTRUMENTATION

*edited by*

JAMES S. MATTSON HARRY B. MARK, JR. HUBERT C. MACDONALD, JR.

*Volume 1:* Computer Fundamentals for Chemists

*Volume 2:* Electrochemistry: Calculations, Simulation, and Instrumentation

*Volume 3:* Spectroscopy and Kinetics

*Volume 4:* Computer-Assisted Instruction in Chemistry (in two parts). Part A: General Approach. Part B: Applications

*Volume 5:* Laboratory Systems and Spectroscopy (in preparation)

# COMPUTER-ASSISTED INSTRUCTION IN CHEMISTRY

*(IN TWO PARTS)*

## Part B: Applications

*EDITED BY*

James S. Mattson

*Division of Chemical Oceanography*
*Rosenstiel School of Marine and Atmospheric Sciences*
*University of Miami*
*Miami, Florida*

Harry B. Mark, Jr.

*Department of Chemistry*
*University of Cincinnati*
*Cincinnati, Ohio*

Hubert C. MacDonald, Jr.

*Koppers Company, Inc.*
*Monroeville, Pennsylvania*

MARCEL DEKKER, INC. New York 1974

MARCEL DEKKER, INC.

305 East 45th Street, New York, New York 10017

LIBRARY OF CONGRESS CATALOG CARD NUMBER: 73-89669
ISBN: 0-8247-6104-9

Current printing (last digit):
10 9 8 7 6 5 4 3 2 1

PRINTED IN THE UNITED STATES OF AMERICA

## INTRODUCTION TO THE SERIES

In the past decade, computer technology and design (both analog and digital) and the development of low cost linear and digital "integrated circuitry" have advanced at an almost unbelievable rate. Thus, computers and quantitative electronic circuitry are now readily available to chemists, physicists, and other scientific groups interested in instrument design. To quote a recent statement of a colleague, "the computer and integrated circuitry are revolutionizing measurement and instrumentation in science." In general, the chemist is just beginning to realize and understand the potential of computer applications to chemical research and quantitative measurement. The basic applications are in the areas of data acquisition and reduction, simulation, and instrumentation (on-line data processing and experimental control in and/or optimization in real time).

At present, a serious time lag exists between the development of electronic computer technology and the practice or application in the physical sciences. Thus, this series aims to bridge this communication gap by presenting comprehensive and instructive chapters on various aspects of the field written by outstanding researchers. By this means, the experience and expertise of these scientists are made available for study and discussion.

It is intended that these volumes will contain articles covering a wide variety of topics written for the nonspecialist but still retaining a scholarly level of treatment. As the series was conceived it was hoped that each volume (with the exception of Volume 1 which is an introductory discussion of basic principles and applications) would be devoted to one subject; for example, electrochemistry, spectroscopy, on-line analytical service systems. This format will be followed wherever possible. It soon became evident, however, that to delay publication of completed manuscripts while waiting to obtain a volume dealing with a single subject would be unfair to not only the authors but, more important, the intended audience. Thus, priority has been given to speed of publication lest the material become dated while awaiting publication. Therefore, some volumes will contain mixed topics.

The editors have also decided that submitted as well as the usual invited contributions will be published in the series. Thus, scientists who have recent developments and advances of potential interest should submit detailed outlines of their proposed contribution to one of the editors for consideration concerning suitability for publication. The articles should be imaginative, critical, and comprehensive, survey topics in the field and/or other fields, and which are written on a high level, that is, satisfying to specialists and nonspecialists alike. Parts of programs can be used in the text to illustrate special procedures and concepts, but, in general, we do not plan to reproduce complete programs themselves, as much of this material is either routine or represents the particular personality of either the author or his computer.

The Editors

# PREFACE

As chemistry is an important part of most other scientific fields, such as biology, medicine, geology, pharmacy, environmental sciences, engineering, and materials, university and college students majoring in these areas are generally required to take one or more years of chemistry courses. Furthermore, large numbers of liberal arts majors elect to take chemistry as the required physical science in their program. Thus, in large universities as many as five thousand students per year will take freshman chemistry and hundreds will continue on into organic, analytical, and physical chemistry. In the small colleges, although the total numbers are smaller, the size of each class and laboratory is generally just as large with respect to student/teacher ratio. As the prospect for increased enrollment grows, while that for significant additional staff positions decreases in general, this ratio will become even more unfavorable in the next ten years in most institutions. Many chemistry departments are beginning to explore the possibilities of computer-assisted instruction (CAI) techniques and methods in an effort to improve the quality of instruction not only in large classes, but also in small classes, to reduce classroom and laboratory costs, reduce time-consuming teacher tasks, such as grading, and to provide special instruction to certain groups of students.

Recently, there have been several conferences, articles in the *Journal of Chemical Education*, etc., which have presented the results of the initial efforts in CAI in chemistry by numerous groups. It was felt that it would be useful and timely to assemble the various methods and opinions on applications of CAI in chemistry in one place, as there are very large numbers of teachers and chemistry departments throughout the country who are considering employing CAI and would benefit from the experiences of pioneers in the area. Thus, we have collected chapters on various aspects of CAI by chemistry teachers from both large and small institutions who have been actively developing and applying CAI techniques. The readers will notice that there is considerable overlap and duplication in the material presented in the different chapters in this volume. We did this purposely as we felt that CAI in chemistry is a developing and evolving teaching technique and that there are no proven methods. We wanted different authors to express their opinions, ideas, approaches, etc., on the same objects and classroom needs. In this way, the reader can evaluate these diverse opinions and experiences in light of his particular situation and, hopefully, arrive at the best approach to the solution of his teaching problems.

This volume is organized into two parts, A and B, for convenience. Part B is subdivided further into two sections. However, the reader should note that, in most cases, portions of chapters in one category also overlap or fall into other categories.

Chapter 1 of Part A serves as a general guide to the philosophy, approaches, and applications of CAI in general. It also serves as an overview to the material contained in the rest of the volume. Part A also contains five chapters which discuss general techniques and applications. Chapter 2 is a detailed presentation of computer-assisted instruction and computer-augmented learning techniques that have been developed and applied to the curriculum at practically all levels at the University of Pittsburgh. Chapter 3 deals with the basic concepts of analog and hybrid computers. Although this chapter is not principally concerned with teaching applications, we felt analog and hybrid computers and computation have great potential for chemistry CAI in simulation of systems and instrumentation. These have been used extensively in engineering sciences and, thus, we felt that it would be very worthwhile including a comprehensive discussion of basic principles in this volume to introduce the reader to the subject. Chapter 4 presents the concept of video projection of teletype output to enable the instructor to use the computer <u>on-line</u> in the classroom. As the volume of chemical literature is expanding at an enormous rate, it is important that computerized information storage and retrieval practice, as discussed in Chapter 5, be introduced to the students (and instructors). One of the questions that readers who are contemplating introducing CAI into their curriculum may have is <u>cost.</u> Chapter 6 gives a brief resume of a cost estimate for various configurations of CAI systems.

Part B, Section 1, which also discusses general approaches, is specifically aimed at presenting detailed discussions of the techniques, programs, and experiences in special applications or courses. Chapter 1 deals with the use of APL language in CAI, and Chapter 2 discusses the use of the PLATO system at the University of Illinois in organic chemistry courses. Chapter 3 is concerned with CAI in physical chemistry courses.

Part B, Section 2 deals with special applications and techniques. Chapter 4 discusses detailed aspects of the use of the computer in generating tests, and Chapter 5 is concerned with the use of a time-shared computer system in CAI. Computer simulation of unknowns is covered in Chapter 6, the use of "canned" programs is discussed in Chapter 7, and Chapter 8 deals with techniques of computerized homework preparation and grading.

The Editors wish to gratefully acknowledge the tremendous editorial help of Dr. Thomas A. Atkinson, Department of Chemistry, Michigan

State University and Dr. Richard L. Ellis, Department of Chemistry, University of Illinois. It would have been impossible to have assembled this volume without their expert comments and advice. We also wish to thank Professor Joseph J. Lagowski for reading the volume and writing the introductory guide and overview (Chapter 1 of Part A) which brings such diverse subjects together. This volume would also have been impossible to complete without the efforts of Bonnie Koran, who produced most of the line drawings for the figures. We also acknowledge the help of many of our colleagues who have contributed helpful comments concerning this volume.

Cincinnati, Ohio
February 1974

J. S. Mattson
H. C. Macdonald, Jr.
H. B. Mark, Jr.

## CONTRIBUTORS TO THIS VOLUME

MORRIS BADER, Department of Chemistry, Moravian College, Bethlehem, Pennsylvania

THOMAS R. DEHNER,* State University of New York at Binghamton, Binghamton, New York

DANIEL B. DONOVAN, Science Department, Corning West High School, Painted Post, New York

JAMES R. GHESQUIERE, The Roger Adams Laboratory, Department of Chemistry, University of Illinois, Urbana, Illinois

JOHN W. MOORE, Department of Chemistry, Eastern Michigan University, Ypsilanti, Michigan

BRUCE E. NORCROSS, State University of New York at Binghamton, Binghamton, New York

FRANKLIN PROSSER, Research Computing Center, Indiana University, Bloomington, Indiana

MANFRED G. REINECKE, Department of Chemistry, Texas Christian University, Fort Worth, Texas

LARRY R. SHERMAN, Department of Chemistry, North Carolina Agricultural and Technical State University, Greensboro, North Carolina

STANLEY G. SMITH, The Roger Adams Laboratory, Department of Chemistry, University of Illinois, Urbana, Illinois

PHILIP E. STEVENSON, Department of Chemistry, Worcester Polytechnic Institute, Worcester, Massachusetts

CLIFFORD G. VENIER, Department of Chemistry, Texas Christian University, Fort Worth, Texas

N. DOYAL YANEY, Department of Chemistry, Calumet Campus Section, Purdue University, Hammond, Indiana

*Present address: Verrazzano College, Saratoga Springs, New York 12866.

CONTENTS

CONTENTS OF PART A

# COMPUTER-ASSISTED INSTRUCTION IN CHEMISTRY

*(IN TWO PARTS)*

**Part B: Applications**

# Section I

# SPECIFIC APPLICATIONS

Chapter 1

# LABORATORY AND CLASSROOM USE OF AN INTERACTIVE TERMINAL LANGUAGE (APL)

Thomas R. Dehner* and Bruce E. Norcross

State University of New York at Binghamton
Binghamton, New York



*Present address: Verrazzano College, Saratoga Springs, New York 12866.

# I. INTRODUCTION

## A. Background

In the past few years we have witnessed a greatly increased effort in many disciplines to realize some of the educational potential presented by the development of computers. Even though considerable progress has been made, further effort and experimentation are needed. A number of different systems and languages are being used for computer-aided instruction (CAI) with varying degrees of success and promise [1-8]. This chapter presents a discussion of some applications of one specific language, APL (A Programming Language [2]), to undergraduate chemistry instruction.

Particular approaches to CAI are as variable as the designed versatility of the modern computer will permit. Our approach came about after the completion of an exploration and itemization of our fundamental eductional philosophies, the difficulties we were having implementing our objectives in large classes, and the ways in which the computer might help to overcome some of these problems. An understanding of these background philosophies and developing problems is essential to understanding our particular computer applications.

SUNY-Binghamton, one of the four university centers of the New York state system, was built upon the established Harpur College, formerly the most selective liberal arts undergraduate college of the SUNY system. Thus, a school with an established tradition of rigorous undergraduate classes, with highly qualified students, a faculty for the most part interested in the undergraduate liberal arts approach to eduction, and a low student/faculty ratio, found itself transformed in less than five years to a university center. The undergraduate enrollment more than tripled, and an active graudate school and graduate faculty were established. Undergraduate admission standards were lowered, and programs for special groups introduced many less qualified students into all class levels. The needs for remedial and specialized and sensitive tutoring in many areas of chemical concepts and in problem-solving approaches increasingly insinuated themselves into class time. In brief, a great deal more teaching time is required to deal with this larger and more heterogeneous student body.

The computer is becoming so important to all advanced work in chemistry, even synthesis [3], that early acquaintance with some aspects of the computer is useful for anyone considering going on in the field. Most students have had some contact with the computer, if only in connection with consumer billing and record keeping, or as a television or other entertainment medium character.

Thus, students are predisposed by their cultural experience to be intrigued by what can be done with a computer. In fact, this predisposition may stimulate interest and accomplishment in the subject matter. These general factors are important, and are ordinarily present in any specific computer application to chemical instruction.

## B. Objectives and Program Types

The identification of firm objectives for CAI includes a consideration of the computing system to be used and the choice of program language. Indeed, the computer system available strongly influences (and sometimes even determines) the choice of language and the specific direction of any CAI effort. While some kinds of CAI objectives can be (and often are) met by a conventional batch-processing system, the dimension of immediate communication added by an interactive system seems to be essential for many CAI uses. A system dedicated to CAI in a particular discipline will include the design and use of hardware and software which is best fitted to the problems and objectives of that discipline. The more common situation, however, is to have available a central computing system that must serve a wide variety of users and applications. In such cases, it is essential to have a language which is flexible and powerful enough to support CAI efforts in many different disciplines. Ideally, the language should be (1) one which is easy to program and easy to use; (2) one which allows direct and immediate communication between the user and the computer; (3) one which permits extended conversational as well as computational operations; and (4) one which accommodates auxilliary interactive facilities, such as slides, tapes, and CRT display. Items 2 and 4 require an interactive terminal system.

We believe that APL has potential as a generally applicable language for CAI in all disciplines. We have experimented with the use of APL in the chemistry curriculum at various levels, from the introductory course for nonscience majors to the graduate course in physical-organic chemistry. Program types include laboratory simulation, laboratory data reduction and analysis, remedial drill programs, and problem-solving tutorials. We have not attempted to redesign our curriculum, or specific courses, around the computer to create a "complete CAI environment" [4], but rather have attempted to integrate the use of computers into existing courses as an additional and unique teaching aid. For our discussion, such applications may be divided into two types with rather different objectives.

Often the first type of CAI effort introduced into a curriculum includes remedial drill programs or extensive tutorial programs. The computer is the impersonal and infinitely patient instructor, allowing the student to progress without embarrassment and at his own pace through particular subject material, all the while wisely questioning the student to test his understanding of the material. Such a program is very useful in a class of widely varied abilities, since a student who finds himself falling behind has recourse to a specialized study aid, to bring his specific skill competence back up to level without constant or embarrassing referral to a faculty member or teaching assistant. The availability of such a program many hours a day, seven days a week, also much more accurately reflects the distribution of student study time than does any ordinary fixed or flexible routine of class, laboratory, and office-hour time.

However, while this type of program is often the first attempted, it is undoubtedly one of the most difficult to write [4]. It has been pointed out previously [6], and is further emphasized here, that any CAI tutorial program should ideally be constructed with sufficient restraint and care to direct students to find their own acceptable answers and procedures, without forcing them into any one and only "correct" answer. This problem of allowing individuality of approach, while still presenting factual material in a palatable way, greatly increases the time and effort "cost" of this type of program.

The second approach to CAI includes programs designed to accept student input and treat it as research data. Such programs may be structured nearly as tightly as the tutorial programs referred to above, or they may be indistinguishable from a typical research program, containing only an I/O format and the algorithm for data treatment. Examples of such programs are (1) programs for data reduction and statistical treatment of experimental data; (2) the simulation program, which allows the student to concentrate on the interdependence of physical variables or to examine parameter relationships in physical models; (3) real-time, on-line uses as a part of a laboratory course; (4) student-written programs. Programs like these are more easily written and debugged than are remedial drills or extensive tutorials. Many specific examples have been reported [1, 7, 8]. We include student-written programs in this major subdivision since they seem to appear simultaneously with student experience with programs of this type. After about one semester of exposure to some formalized computer contact, a significant fraction of the students begin to use the computer routinely for laboratory and course-related data reduction and analysis. The extent to which this occurs is controlled in large part by the availability of the computer facilities and the cooperation of the computer staff.

Our own experience with CAI began initially with a few students in one of the second semester freshman chemistry recitation sections using programs of the second type. These students worked with two titration programs written in APL by Science Research Associates (Chicago, Illinois), and somewhat modified by us. These programs allowed the students to simulate acid-base titrations for a variety of weak acids, and turbidimetric titrations to obtain solubility product information for a number of slightly soluble salts. Response from the limited number of students involved with these programs was enthusiastic. In fact several volunteers were coming at 7:00 a.m. in order to use the computer. Although this type of program does not provide experience with lab technique, it can help the student grasp the basic concepts involved in titrimetric analysis, and thus make his lab time more meaningful.

Later in the year a set of simple kinetics programs introduced the students to statistical treatment of experimental data. These kinetics programs apply first- and second-order treatments to the data, calculate the best straight-line fit by a simple least-squares routine, and print out the rate constant calculated from the slope, and the standard deviation. The student can choose to see both first- and second-order plots, and can compare the rate constant value from the least-squares treatment to the one he has obtained from his own hand-drawn graph of the same data. An additional program was added to this set for the calculation of the activation energy from the temperature dependence of the rate constant.

APL programs have also been used in some of the advanced courses in the chemistry curriculum. For example, in the organic laboratory a program is used to calculate the percent composition of liquid mixtures from vapor phase chromatography data. For the introductory physical chemistry course a program has been written which calculates the eigenvalues and wavefunctions for a one-sided potential well. By varying the parameters (depth and width of well and mass of particle) the student can follow the effect of parameter variations on the eigenvalues and attempt to discover valid relationships. The program for determination of activation energy, mentioned earlier, was actually developed for use by juniors and seniors in a physical chemistry lab experiment on the internal rotation of amides, measured by a NMR method. The data obtained are not susceptible to simple least-squares treatment, so the students used a computer program to calculate weighting functions for statistical treatment of their NMR data. In the physical-organic courses, students have been required to develop their own line-fitting program in APL for treatment of kinetics and other experimental data.

It is not possible to discuss in detail each of the program types we have tried; several of these have been described in detail elsewhere [7]. Instead, we attempt, by discussing a specific program and its development in considerable detail, to illustrate one way in which APL can be used effectively by a large class in general chemistry. We hope the discussion which follows will convey to the reader a sense of the value of such an effort in CAI, an awareness of some of the problems encountered and possible solutions, and some appreciation for both the utility and the limitations of APL as a language for CAI.

## II. SYSTEM CONSIDERATIONS

### A. Language

First developed and used at IBM in the early 1960s, APL is an interactive, algorithm-oriented, time-sharing language with a relatively simple notation and format accommodating both computational and conversational applications. By means of the typewriter terminal the user can define and store programs and execute system commands or mathematical statements, and receive an almost immediate response from the computer. APL is simultaneously an easy language to learn and use and a sophisticated language with great computational power--not a paradox since it is a function of the involvement and applications chosen. APL notation is similar to familiar algebraic notation. The primitive functions and operations in APL offer great flexibility and simplicity in mathematical and logical operations, particularly in manipulation of vectors and matrices.

Examples of the power of some of the defined functions, part of the APL language, are illustrated in Fig. 1-3. Figure 1 demonstrates the usefulness of "plus reduction", (+/X), which sums the elements of the vector, or string of numbers, "X". The symbol ρ, rho, gives the number of components in the vector. The average, then, is calculated quite simply by dividing the sum of the elements by the number of elements. Figure 2 shows the use of another function, ⍋ (an overstruct character, the "grade-up"), which in one step returns a new, rearranged vector, listing the same numbers in order of increasing magnitude. Figure 3 demonstrates some of the power of APL in dealing with matrices; in this case, the solution of three linear equations in three unknowns. Both the vector of constants (B) and the coefficient matrix (A) are defined from the equations, and then the solution is obtained in one step using the APL primitive, "quad-divide", (⌹), which performs the appropriate matrix algebra.

Not only can the language handle fairly sophisticated calculations and mathematical operations, but it also allows programming in the extended conversational mode so that instructions may be given, and questions asked, in English prose using character vectors as well as in mathematical formulations. Such programming, and the results on execution of the program, are shown in Fig. 4.

| BASIC | FORTRAN | APL |
|---|---|---|
| 10 DIM X (100) | DIMENSION X (100) | (+/X)÷ρX←⎕ |
| 20 LET S=0 | READ (t,10) N, (X(I),I=1,N) | |
| 30 READ N | 10 FORMAT (15, (E15.2)) | |
| 40 FOR I=1 TO N | S=0.0 | |
| 50 READ X (I) | DO 9, I=1,N | (Typing this line at a terminal would |
| 60 LET S=S+X(I) | 9 S=S+X(I) | result in a request by the computer for |
| 70 NEXT I | A=S/N | a list of numbers (⎕:). After such a |
| 80 LET A=S/N | WRITE (6,20)A | list has been entered, the computer |
| 90 PRINT A | 20 FORMAT (E15.2) | would return the desired average.) |
| 100 DATA | END | |
| - - - - - | | |
| - - - - - | | |
| XXX END | | |

Fig. 1. Averaging a set of numbers.

BASIC

```
 10 DIM X (100), Y(100)
 20 READ N
 30 FOR I=1 TO N
 40 READ X (I)
 50 NEXT I
 60 FOR I=1 TO N
 70 LET A=X(I)
 80 LET L=1
 90 FOR J=1 TO N
100 IF A=X (J) THEN 130
110 LET A=X (J)
120 LET L=J
130 NEXT J
140 LET Y (I)=A
150 LET X(L)=100000.
160 PRINT Y(I)
170 NEXT I
180 DATA
     - - - - -
XXX END
```

FORTRAN

```
   DIMENSION X (100), Y(100)
   READ (5,10) N, (X(I),I=1,N)
10 FORMAT (15, (E15.2))
   DO 9 I=1,N
   A=X(I)
   L=1
   DO 8 J=1,N
   IF (A-X(J)) 8, 8, 7
 7 A=X(J)
   L=J
 8 CONTINUE
   Y (I)=A
 9 X (L)=100000
   WRITE (6,20) (Y(I),I=1,N)
20 FORMAT (E15.2)
   END
```

APL

```
X[⍋X←⎕]
```

(Typing this line at a terminal would result in a request by the computer for a list of numbers (⎕:). After such a list has been entered, the computer would return the ordered vector.)

Fig. 2. Sorting a set of numbers.

| Equations | APL |
|---|---|
| $5X_1 + 2X_2 + X_3 = 36$ | `B← 36 63 81` |
| $X_1 + 7X_2 + 3X_3 = 63$ | `A← 3 3ρ5 2 1 1 7 3 2 3 8` |
| $2X_1 + 3X_2 + 8X_3 = 81$ | `B⌹A` |

```
3.6   5.4   7.2
```

Fig. 3. Matrix solution of simultaneous equations.

Finally, programs can be edited and modified easily, and can be stored in public libraries for ready access by any user. Brief handout instructions, supplemented by individual instruction at the terminal generally supplied by student proctors, is sufficient to allow introductory students to use the programs made available. At SUNY-Binghamton, interested students can attend a set of video-taped lectures on APL which are shown several times a year. A student or faculty member who has attended only a few of these lectures can begin to write his own programs. Obtaining increased sophistication in programming is furthered by a number of good APL texts [9] and the enthusiastic collaboration of APL users on campus. Furthermore, one of the most attractive features of APL is that a program need not be polished in order to work well. This fact is demonstrated by two programs shown in Fig. 4. Both successfully convert Centigrade temperatures to Fahrenheit. Both require the input of the same data. The second accomplishes in four lines of programming the same operations accomplished in ten lines in the first program. Even though the first program was written by a newcomer to the language, it works quite satisfactorily.

Although APL has many features which make it attractive for CAI, there are at the present time unfortunate restrictions on its flexibility. Full implementation of APL/360 requires a relatively large computer (128K-170K bytes). The system is also presently rather weak in its limited support of such auxilliary facilities as rapid I/O devices, file capability, workspace catenation capability, cathode ray tube (CRT), graphic display, and operation of slide projectors or tape recorders. Some of the major problems associated with APL CAI arise directly from this lack of support for peripherals and the need for an ability to expand or catenate workspaces. For instance, the display of organic structural formulae to students, done excellently by the dedicated language PLATO [5], is not possible at present with APL. An alternative to this approach is to use a slide projector, as in certain COURSEWRITER routines. Graphic display of data by the APL plot function on a terminal is very slow and limited in resolution. The availability of

PROGRAM

```
     ∇ TVERT
[1]    'INTERVAL?'
[2]    I←⎕
[3]    'BEGINNING OF RANGE?'
[4]    C←⎕
[5]    'END OF RANGE?'
[6]    E←⎕
[7]    F←32+1.8×C
[8]    'CENTIGRADE:';C;' FAHRENHEIT:';F
[9]    C←C+I
[10]   →(C≤E)/7
     ∇
```

EXECUTION

```
          TVERT
INTERVAL?
⎕:
          8
BEGINNING OF RANGE?
⎕:
          ¯40
END OF RANGE?
⎕:
          8
CENTIGRADE:¯40 FAHRENHEIT:¯40
CENTIGRADE:¯32 FAHRENHEIT:¯25.6
CENTIGRADE:¯24 FAHRENHEIT:¯11.2
CENTIGRADE:¯16 FAHRENHEIT:3.2
CENTIGRADE:¯8 FAHRENHEIT:17.6
CENTIGRADE:0 FAHRENHEIT:32
CENTIGRADE:8 FAHRENHEIT:46.4
```

PROGRAM

```
     ∇ VTVERT
[1]    'ENTER BEGINNING OF RANGE, INTERVAL,
       AND NUMBER OF VALUES...IN THAT ORDER.
       '
[2]    IP←⎕
[3]    'CENTIGRADE    FAHRENHEIT  '
[4]    ((IP[3],1)ρC),(IP[3],1)ρF←32+
       1.8×C←IP[1]+IP[2]×¯1+ιIP[3]
     ∇
```

EXECUTION

```
          VTVERT
ENTER BEGINNING OF RANGE, INTERVAL, AND
          NUMBER OF VALUES...IN THAT ORDER.
⎕:
          ¯40 8 8
CENTIGRADE    FAHRENHEIT
    ¯40              ¯40
    ¯32              ¯25.6
    ¯24              ¯11.2
    ¯16                3.2
     ¯8               17.6
      0               32
      8               46.4
     16               60.8
```

Fig. 4. Temperature interconversions.

CRT support would permit a much more rapid and effective display of function and parameter relationships best visualized and understood in graphic form. These capabilities, not readily available in APL now, are allegedly nearing release.

Other serious limitations are the fixed size of the workspace and the lack of file capability. If a relatively complex program is attempted, which involves storage of student names and performance data, a limit on the number of students accommodated may be reached at anywhere from about ten to twenty-five students, depending on the program complexity and kinds of data stored. In the application of CHEMLAB, we are able to store only the names and data of thirteen students per workspace.* An ability to call files stored outside the workspace, or to expand the workspace size by linking on one or more additional workspace units, would greatly ease this problem.

The need for these additional features of APL has been called to the attention of the developers of the language [10], and assurances have been given that developments are pending in these areas. File capability has recently become commercially available (APL PLUS, Scientific Time Sharing Corp., a subsidiary of I. P. Sharp, Toronto, Canada). Additional features, such as an extended semantics including some new defined functions and a limited graphics APL character set, are being explored.

Another reservation noted in adopting APL as a CAI language was the limited transportability of the programs. In 1967 SUNY-Binghamton was the first non-IBM site in this country to have APL installed as a resident system. However, by the end of 1971 there were over 30 educational institutions in the United States operating the IBM-APL system. In addition, most other major computer companies are in the process of developing their own versions of APL. Thus, it appears that limited transportability is no longer a serious handicap.

---

* The main function, CHEMLAB1, is approximately 260 steps and requires about 15,000 bytes. Supplemental functions used by CHEMLAB1 (FIND, VS, AND, SPACE, DFT, and CHECKR) total about 85 steps, and require an additional 5,000 bytes. Stored correction tables and variables require approximately 3,500 bytes. Student name and data matrices require 175 bytes per student. Since another approximately 5,000 bytes are needed during execution, this means only approximately 15 students can use a given 32K byte workspace. It requires two workspaces to service the twenty-four students in one laboratory section (this is not a serious drawback in this particular case, since ordinarily two terminals, and thus two workspaces, are used per laboratory section).

## B. Support

Initially we were using an IBM 360/40 computer with a minimal memory and a small number of terminals. APL was available only three hours per day. Consequently, only a fairly small number of students were involved in the use of our programs. It is interesting to note that the terminals were in use 90% of the time when APL was available, compared to approximately 10% use during remote FORTRAN (RAX) hours. Recently we acquired an IBM S/370/155 computer, and APL is now available on campus ten hours per day. Currently we have thrity-three remote typewriter terminals (#2741) on campus. In addition, 15-20 terminals in area high schools and regional colleges are tied to our system. All students in general chemistry (over 300) use the computer as part of their class assignment.

We have found two kinds of terminal availability important, reflecting the different uses involved. For general purposes, a group of terminals in one room, under the control of the computer center, has a number of advantages. A computer terminal is a significant investment deserving some protection and control; it requires continuous simple but essential maintenance, such as ribbon and paper change, type-ball exchange, impact-pressure adjustment, and transmission line adjustments. None of these mechanical details is excessively complicated, but our experience suggests that when dealing with large number of initially naive participants, it is well to have full-time proctors in charge of the terminals. The cost of such proctoring per contact hour falls when terminals are centralized. Nevertheless, several satellite groupings of terminals may be useful, particularly if concentrations of users may be identified. From time to time we have had four to eight terminals in a room in the Physical Science Building, and the rest, with only a few special exceptions, in one central location.

Another way in which we use terminals is in connection with data generated during regularly scheduled laboratories. For this use we connected terminals to transmission line branches in a room convenient to the freshman chemistry labs (Fig. 5). This allowed students to use a program for treatment of laboratory data during their regularly scheduled laboratory period.

While it is certainly true that a chemistry faculty member can do a more than adequate job of programming and administering the use of an APL CAI unit, access to one full-time programmer is very helpful, greatly increasing the possible program output. We have also found that hiring experienced students as terminal proctors increases the efficiency of the terminal usage over an unattended situation. In a typical lower-division undergraduate laboratory situation we have tried both systems: terminals in a room adjacent to the laboratories, but proctored only by the faculty and teaching assistants ordinarily scheduled in the laboratories; and the same physical situation but

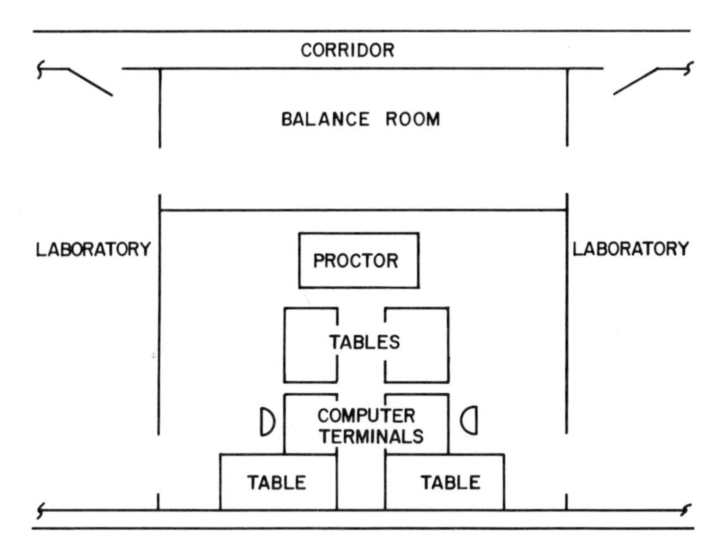


Fig. 5. Freshman laboratories and the interactive terminals.

with the addition of one additional staff member assigned exclusively to the terminal rooms. The second system proved to be an unexpectedly great improvement over the first, and is highly recommended.

## C. Representative Costs

The major costs in a CAI program are terminal rental, student-generated terminal time, and programming time. Cost calculations are difficult, and depend heavily on the specific items considered; therefore, cost comparisons between systems should be made with caution.

For our system, terminal rent is about $1100 per year per terminal. Connect time may be estimated at a current rate of $3.00 per hour. A data-set connection costs about $13 per month. A Tee connection, which allows one terminal to be used in either of two positions, adds about another $2 per month per tee. An alternative to the data set is a four-wire, double-circuit, leased line, for which the charge is established on a mileage basis. Such a line over a distance of 0.2 miles costs about $9.00 per month, with Tee. Providing a student proctor for 40 hours a week at a minimum wage (a condition sometimes required by a computer center in exchange for allowing satellite grouping of terminals) may add another $300 per month. A typical calculation (assuming four terminals, an average student connect-time of an hour a week) yields an estimated cost of $20,000 per nine-month year for a class of 100 students, or roughly $200 per student, of which $120/

student is terminal time. This is about $5 per student-hour of connect time. These costs are often borne to a greater or less degree by the computer center rather than a specific department, and may be hidden in lump-sum equipment budgets. They do, however, reflect a fairly accurate cost figure, and should be kept in mind if one is considering a new installation.

The faculty investment in programming time is difficult to determine since programs undergo almost continual evolution even after considerable use. Conservative estimates from our experience suggest that the initial development and debugging of a long tutorial or involved data treatment program like CHEMLAB requires about 100 hours of full time effort, while shorter, simple problem-solving programs require only a few hours. In addition, programs for general class use during specified hours (like CHEMLAB) require a few hours before the lab week for organizing and arranging workspaces and schedules with the computer center.

## III. CHEMLAB

The program CHEMLAB was a response to the increasing freshman chemistry enrollment pressure on the effectiveness of one of our intellectually more demanding introductory chemistry experiments. This standard experiment, the molar volume of gases, appears in many general chemistry laboratory manuals.

At a time when we were seriously considering dropping this experiment for some of the nonacademic reasons alluded to in the introduction, we heard a paper presented at the Symposium on the Status of Computer-Assisted-Instruction in Chemistry held at Toronto in May, 1970 [11]. In this paper D. S. Olson and colleagues described the use of computer punch cards for grading chemistry laboratory reports for large classes at the Air Force Academy. We realized that by using an interactive system such as APL, and by placing computer terminals in the laboratory, we could not only ease the grading burden for our teaching assistants and collect data for class analysis, but, more importantly, we could also permit the student to identify possible experimental error while his apparatus was still available for repeats. To understand the development of this program, we need to describe in more detail the experiment we chose for our first attempt at this kind of computer interaction.

### A. The Experiment

The purpose of the molar volume experiment is to expose the student to quantitative techniques in working with gases and to verify Avogadro's hypothesis concerning gas behavior. The student weighs out a sample of a mixture of potassium chloride (KCl) and potassium chlorate ($KClO_3$) of unknown percent composition and, with appropriate apparatus which he has assembled, decomposes the $KClO_3$ by heating and collects and measures the oxygen gas evolved. From the primary data collected (weights, volume of water displaced, temperature, barometer reading) the student can then calculate a molar volume for oxygen at standard temperature and pressure and the percent $KClO_3$ in his sample.

The laboratory procedure is a fairly demanding one for students at this level, and there is plenty of opportunity for experimental error. Six pieces of primary data must be obtained. On these the student performs a number of calculations, including aqueous vapor pressure and barometer scale expansion corrections, before he arrives at his final answer. Typically, a student might carry out one determination and just finish his calculations by the end of the lab period, only to find that his final answer is unacceptable and he must repeat the experiment. Clearly, it would be helpful to the student to have an immediate check on his primary data.

## B. The Interactive Program

The objectives of the computer program CHEMLAB are: (1) to allow the student to check immediately his experimental measurement of the volume of oxygen and to identify the error range of his measurement for his sample and experimental conditions; (2) to allow the student to check his calculations of molar volume and percent composition for his particular sample; (3) to store student primary data and results for subsequent analysis and grading; (4) to encourage the student to attempt an intelligent discussion and analysis of his own experimental results.

As soon as the student has finished collecting data for his first run, he goes to the computer terminal in the lab and types CHEMLAB1, which starts program execution. (Just prior to the laboratory period the lab instructor has signed on the computer and loaded a copy of the CHEMLAB workspace from the public library into his active workspace by a simple system command.) The interaction of student with computer program is shown diagrammatically in Fig. 6.

The introductory steps in the program establish initial values of needed variables and counters (CHEMLAB1, steps 1-8, see Appendix). The program next asks whether or not the student needs help in using the program. If help is requested, the detailed instructions contained in lines 13-25 of the CHEMLAB1 listing are printed out. These are also a part of the regular laboratory instructions.

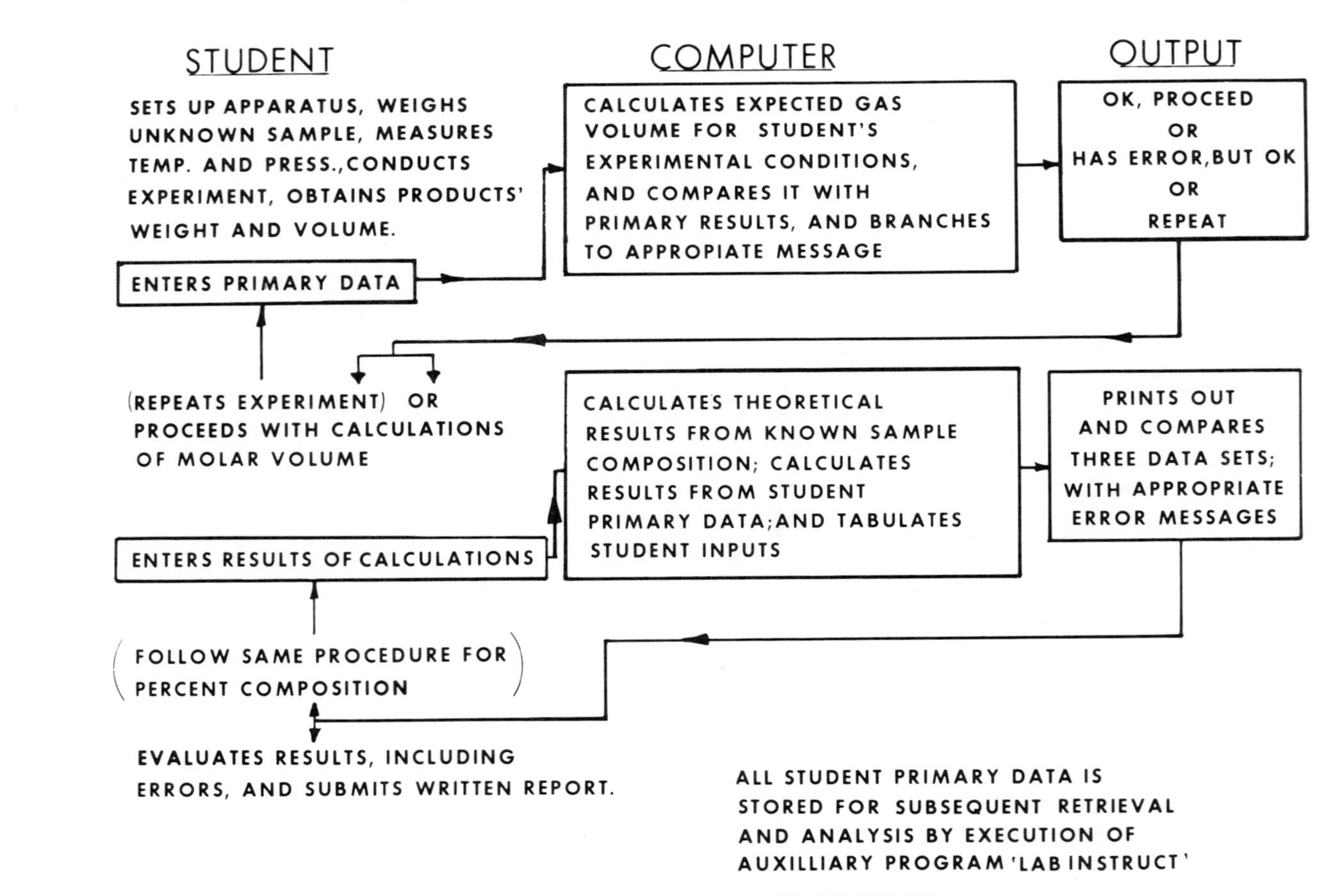


Fig. 6. Student interaction with CHEMLAB.

The program has three main parts for data entry. Steps 27-36 of CHEMLAB1 branch the student to that part of the program appropriate to his progress. This is done by means of a counter advanced for each student as he successfully completes each part. Two sets of data may be entered--trials 1 and 2.

Part 1 of CHEMLAB1 (steps 37-127) provides for entry and checks of primary experimental data. The student is instructed to enter his data, which includes weights, temperature, barometer reading, and collected volume of oxygen. The program then calculates the theoretical volume of oxygen and theoretical weight loss (using student's conditions of temperature and pressure, and using percent composition information stored in program for the designated sample) and compares these with the student's experimental values. The computer response is determined by the agreement between experimental and theoretical values. The agreement between the experimental gas volume found by the student and that volume calculated by the computer determine the level of acceptability at this point. If the difference between the student volume and calculated volume is less than 10%, then the message returned is "very good". A value between 10 and 20% yields "reasonably good". Between 20 and 30% results in "acceptable....but poor", and over 30% is "unacceptable". These tests are contained in CHEMLAB1, line 90. The limits of acceptability are easily modified by changing the ranges in this step. (A similar check is made in steps 116-125 for weight loss.) If the student's data is acceptable, he is told to proceed with his calculations. His name and data are stored in the workspace and the sequencing counter is advanced, so that the next time the same student calls for the program he is automatically in Part 2. If the agreement is fairly good, the student is allowed to proceed with his calculations, but is advised to repeat the experiment. If the agreement is poor, then the student is told that his data is unacceptable and that he must repeat the experiment. In the case of fair or poor results, leading questions are asked by the computer to help the student pinpoint likely sources of error. These features are illustrated in sample executions of Part 1, in Section III.C of this chapter.

After the student calculates from his primary data the weight of oxygen, the partial pressure of oxygen, and the molar volume of oxygen at STP, he returns to the terminal and types CHEMLAB1. Upon entry of his name and trial number, he is automatically branched to Part 2 (Appendix, steps 128-214) which provides for entry and check of his calculated values. Then the student enters his calculated values. The program then does the same calculations, using the student's primary data which is stored in the DATA array and the correction factors stored in the workspace. The computer also calculates the theoretical values for the particular sample which the student has analyzed. The three columns of data are then printed out side by side for comparison (see sample execution of Part 2 in Section III.C). This print-out allows the student to check immediately (a) the validity of his calculations, and (b) the accuracy of his experimental measurements. The student includes this part of the computer printout in his lab report, and is expected to discuss the agreement or disagreement of the three columns. In many cases, because he can check his own errors, a more intelligent discussion of his experimental results is possible.

From the experimental weight of oxygen, the known molecular weight of $O_2$, the total sample weight, and the balanced equation for the reaction, the student can calculate the percent potassium chlorate in his sample. He then returns to the terminal, is branched to Part 3 (Appendix, steps 215-269) and enters the results of this calculation. The program does the same calculations and prints out the two columns of data, along with the theoretical percent $KClO_3$ for the student's sample (see sample execution, Part 3, section III.C). The student can check both the validity of his calculations and the accuracy of his results.

Once data is accepted and saved by the computer, it is stored in the workspace and cannot be altered. Another program, LABINSTRUCT, permits the laboratory instructor to retrieve all of the names and data from his lab section in a defined format for comparison and analysis. If calculations in Part 2 or Part 3 are in error, the student must show the corrected calculations as part of the final laboratory report. This feature prevents the student from misusing the computer-calculated results.

## C. Sample Executions

### 1. Illustrating Unacceptable Primary Data--Part 1 Execution

```
      CHEMLAB1


PLEASE ENTER YOUR NAME, FIRST NAME FOLLOWED BY LAST.
JACK RACK
OK JACK LET'S GET ON WITH IT!

       DO YOU REQUIRE ASSISTANCE ? (TYPE YES OR NO)
⎕:
      NO

       TRIAL ? (TYPE 1 OR 2)
⎕:
      1
       SAMPLE DESIGNATION ? (IF PURE KCLO3, TYPE PURE)
⎕:
      A
  PART 1:  ENTER THE FOLLOWING PIECES OF DATA:

|1|   WEIGHT OF TUBE AND CONTENTS BEFORE HEATING.
⎕:
      35.2347
|2|   WEIGHT OF TUBE AND RESIDUE AFTER HEATING.
⎕:
      34.9161
|3|   WEIGHT OF EMPTY TUBE.
⎕:
      31.7249

|4|   TEMPERATURE OF THE OXYGEN [CENTIGRADE]:
⎕:
      24.0
```

```
|5|    BAROMETRIC READING; UNCORRECTED AT ROOM TEMPERATURE.
⎕:
       742.0
|6|    VOLUME (UNCORRECTED) OF OXYGEN COLLECTED.
⎕:
       125

       HAVE YOU MADE ANY ERRORS IN ENTERING YOUR DATA ?
⎕:
       NO

        YOUR EXPERIMENTAL VOLUME IS UNACCEPTABLE (ERROR>30PERCENT). PLEASE REPEAT THE
  EXPERIMENT.  YOU WILL BE ALLOWED TO ENTER YOUR NEW DATA UNDER THE SAME TRIAL NUMBER.
        YOUR MEASURED VOLUME IS TOO LOW. CONSIDER THE FOLLOWING POSSIBLE SOURCES OF ERROR:

             DID YOU ENTER WEIGHTS CORRECTLY?
             DID YOU COMPLETELY DECOMPOSE THE KCLO3 IN YOUR SAMPLE?  (DID YOU HEAT LONG
      ENOUGH?)
             DID YOU HAVE AN AIRTIGHT SYSTEM (NO LEAKS!)?
```

## 2. Illustrating Very Good Primary Data and Some Correction Capabilities of the Program--Part 1 Execution

```
        CHEMLAB1


PLEASE ENTER YOUR NAME, FIRST NAME FOLLOWED BY LAST.
KEVIN KELLEY
OK KEVIN LET'S GET ON WITH IT!

         DO YOU REQUIRE ASSISTANCE ? (TYPE YES OR NO)
⎕:
        NO

         TRIAL ? (TYPE 1 OR 2)
⎕:
        1
         SAMPLE DESIGNATION ? (IF PURE KCLO3, TYPE PURE)
⎕:
        A
  PART 1:  ENTER THE FOLLOWING PIECES OF DATA:

|1|     WEIGHT OF TUBE AND CONTENTS BEFORE HEATING.
⎕:
        35.2347
|2|     WEIGHT OF TUBE AND RESIDUE AFTER HEATING.
⎕:
        34.9161
|3|     WEIGHT OF EMPTY TUBE.
⎕:
        31.7249
|4|     TEMPERATURE OF THE OXYGEN [CENTIGRADE]:
⎕:
        34.0
        DATA OUTSIDE RANGE OF TEMPERATURE CORRECTION TABLE; 16≤T≤33 °C
|4|     TEMPERATURE OF THE OXYGEN [CENTIGRADE]:
⎕:
        24.0
```

```
|5|    BAROMETRIC READING; UNCORRECTED AT ROOM TEMPERATURE.
⎕:
       720.2
|6|    VOLUME (UNCORRECTED) OF OXYGEN COLLECTED.
⎕:
       250.0

       HAVE YOU MADE ANY ERRORS IN ENTERING YOUR DATA ?
⎕:
       YES

       ENTER THE NUMBER OR NUMBERS ( FOUND IN THE | | ) OF THE INPUT STATEMENTS.ASSOCIATE
       D WITH
   THE DATA YOU WISH TO CORRECT.  IF MORE THAN ONE NUMBER IS TO BE ENTERED, SEPARATE WITH
        BLANKS.
⎕:
       5
|5|    BAROMETRIC READING; UNCORRECTED AT ROOM TEMPERATURE.
⎕:
       742.0

       HAVE YOU MADE ANY ERRORS IN ENTERING YOUR DATA ?
⎕:
       NO

        YOUR EXPERIMENTAL VOLUME IS IN VERY GOOD AGREEMENT WITH THE THEORETICAL VOLUME(UN
       CORRECTED)
   FOR YOUR SAMPLE AND YOUR EXPERIMENTAL CONDITIONS. PROCEED WITH YOUR CALCULATIONS.
      WHEN YOU NEED ME AGAIN, JUST CALL (CHEMLAB1)!
```

## 3. Illustrating Part 2 Execution

```
      CHEMLAB1

PLEASE ENTER YOUR NAME, FIRST NAME FOLLOWED BY LAST.
KEVIN KELLEY
OK KEVIN LET'S GET ON WITH IT!

       TRIAL ? (TYPE 1 OR 2)
⎕:
      1
 PART 2:   ENTER THE FOLLOWING PIECES OF DATA:
|1|   WEIGHT OF OXYGEN COLLECTED.
⎕:
      .3186
|2|   TEMPERATURE, ABSOLUTE.
⎕:
      297
|3|   BAROMETRIC PRESSURE, CORRECTED
              ( TOTAL GAS PRESSURE )
⎕:
      739.3
|4|   PRESSURE DUE TO OXYGEN
⎕:
      716.9
|5|   VOLUME OF OXYGEN AT STP
⎕:
      219
|6|   MOLAR VOLUME OF OXYGEN ( IN LITERS )
⎕:
      22200
MOLAR VOLUME IN LITERS, PLEASE!
|6|   MOLAR VOLUME OF OXYGEN ( IN LITERS )
⎕:
      22.2

      HAVE YOU MADE ANY ERRORS IN ENTERING YOUR DATA ?
⎕:
      NO
```

```
                                              |--------MACHINE COMPUTED
                                              |        USING YOUR DATA.
                              YOUR            |
                              DATA.           |        MACHINE COMPUTED
                                              |    |-- THEORETICAL
                               |              |    |   VALUES.
                               ↓              ↓    ↓

WEIGHT OF OXYGEN.              0.3186*        0.3186   0.3437
ABSOLUTE TEMPERATURE.        297.0          297.2    297.2
CORRECTED BAROMETRIC
    READING.                 739.3          739.1    739.1
PRESSURE DUE TO OXYGEN.      716.9          716.7    716.7
VOLUME OF O2 AT STP.         219*           217      241
MOLAR VOLUME OF O2.           22.2           21.8     22.4

  COMPARISON OF THESE COLUMS ALLOWS YOU TO CHECK BOTH YOUR CALCULATIONS
  AND THE AGREEMENT BETWEEN YOUR EXPERIMENTAL DATA AND THE THEORETICAL
  VALUES FOR YOUR SAMPLE.

    ERROR MESSAGE INTERPRETATION:
            *   5  -  10 PERCENT ERROR
           **  10  -  20 PERCENT ERROR
          ***  20  -  40 PERCENT ERROR
            ≠        >40 PERCENT ERROR

            PLEASE HAND IN THIS TABLE WITH YOUR FINAL LAB
   REPORT.  IF YOUR SAMPLE IS AN UNKNOWN, YOU MAY PROCEED
   TO PART 3  WHEN READY  BY TYPING CHEMLAB1 AGAIN.
```

## 4. Illustrating Part 3 Execution

```
      CHEMLAB1


PLEASE ENTER YOUR NAME, FIRST NAME FOLLOWED BY LAST.
K KELLEY
OK KEVIN LET'S GET ON WITH IT!

       TRIAL ? (TYPE 1 OR 2)
⎕:
      1
  PART 3:  ENTER THE FOLLOWING DATA:
|1|   MOLES OF OXYGEN.
⎕:
      .0100
|2|   MOLES OF POTASSIUM CHLORATE DECOMPOSED.
⎕:
      .0066
|3|   WEIGHT OF POTASSIUM CHLORATE IN SAMPLE.
⎕:
      .823
|4|   WEIGHT OF SAMPLE USED.
⎕:
      3.501
|5|   PERCENT OF POTASSIUM CHLORATE IN SAMPLE.
⎕:
      .25
I WANT A TRUE PERCENT; NOT A FRACTION !
|5|   PERCENT OF POTASSIUM CHLORATE IN SAMPLE.
⎕:
      25

      HAVE YOU MADE ANY ERRORS IN ENTERING YOUR DATA ?
⎕:
      NO
```

```
                                                |----- MACHINE COMPUTED
                                YOUR            |      USING YOUR DATA.
                                DATA.           |
                                  ↓             ↓

MOLES OF OXYGEN EVOLVED.         0.01000       0.00996
MOLES OF KCLO3 DECOMPOSED.       0.00660       0.00664
WEIGHT OF KCLO3 IN SAMPLE.       0.823         0.814
WEIGHT OF SAMPLE USED.           3.501         3.510
PERCENT KCLO3 IN SAMPLE.        25.0          23.2


    THEORETICAL PERCENT POSTASSIUM CHLORATE FOR YOUR SAMPLE IS 25
PERCENT.  YOU HAVE FINISHED COMPUTER TREATMENT OF DATA FOR THIS
SAMPLE, AND MAY NOT ALTER ANY OF THE DATA.  PLEASE HAND IN THIS
TABLE  WITH YOUR REPORT.  IF YOU HAVE MADE ANY ERRORS IN YOUR
FINAL CALCULATIONS, CORRECT THEM AND INCLUDE THE CORRECTIONS IN
YOUR LAB REPORT.  YOUR REPORT SHOULD INCLUDE A DISCUSSION OF
YOUR EXPERIMENTAL ERROR
```

## IV. INTERACTIVE PROGRAM PROBLEMS

It is not the purpose of this section to teach APL, but rather to point out some of the problems we encountered in the interactions between program and student, and how we solved or evaded them. We would expect these observations to be useful in working with any interactive language.

### A. General Suggestions

A major program evolves, in general, from many drafts. Problems often seem to arise where no difficulties were anticipated. It is very important that the academic directions and decisions come from the faculty member. It follows from this that a faculty member needs to keep in close touch with the development and debugging of the program, particularly in the early stages. A good, operating interactive program represents a large investment of programmer and faculty time. Some of this investment may be recovered if an attempt is made to generalize the program so that significant blocks of it can be used in other applications with minor changes.

If the computer is not under your personal control, it is important to consult with the computer center before trying to process a large number of students in a fixed period of time through an interactive program. Languages such as APL are often run simultaneously with other remote languages, and with background batch processing. A low-priority assignment to the system you are using may result in relatively long delays between transmission of a line by a student and the computer response. Delays of 10 sec are long, and of 30 sec or more destructively frustrating. Often a change in priority will reduce such delays to a few seconds at most.

Some students view any kind of computer mediation of instruction as a negative, depersonalizing, undesirable interference with the educational process. It appears to us to be important to emphasize whenever possible that programs such as CHEMLAB are intended to improve student-faculty contact by moving some routine data manipulations to the computer, and using that time saved for dealing with whatever some of the current "real" problems are. In other words, the time spent by the faculty and student in the laboratory is not necessarily reduced, but the questions and discussions which occur seem to be devoted more to "How" and "Why" rather that "Is this number right?" or "What did I do wrong?"

### B. Some Programming Features

We have tried two kinds of student name entry for CHEMLAB. In the more general program, a student enters his name on request, and it is then stored, exactly as written, for future use. In the second version, a name table is filled initially with names of students in a particular laboratory section. These students must then enter that workspace for their computer experience. Both versions are in use, and have particular advantages for different purposes. The first method makes the program generally available to anyone. The second version is more efficient for large-class production use.

Two kinds of data entry are available for this program. Originally a step-by-step method was used (see sample executions, Section III.C), in which each separate item was called for individually. Such entry required a transmission/response sequence for each datum. Since it was important to reduce the time at the terminal for each student during the first entry into the program, which occurs during a laboratory period, an attempt to cut the number of separate entries per student was explored. The approach used was to have the students enter the primary data as a vector (Fig. 7). A clear difference in time for primary data entry was found only when the students prepared for the vector entry by organizing their raw data in the order required before signing on the terminal. This feature can be useful in encouraging students to set up a proper data table in their lab notebooks. Otherwise, the directed nature of the step-by-step entry was more efficient. Under ordinary circumstances, a student should be at the terminal for less than five minutes for each part of CHEMLAB. Two terminals available for the last hour of a three-hour lab would accommodate one section of twenty-four students without excessive delays. Two terminals available for the last two hours of such a three-hour laboratory period would accommodate most of the students in two twenty-four student laboratory sections.

Error messages designed to correct faulty entry seem to inspire some programmers to excesses of cuteness or sarcasm. While a light touch with prose responses is often stimulating and encouraging to the students, it is easy to misjudge the appropriateness of a response from consideration of only one situation. For instance, an early draft error message received by one who entered a quantity requested to be in liters with a five-digit number, obviously milliliters instead, was faced with the message:

I SAID LITERS, DUMMY. TRY AGAIN.

This message might be an appropriate rap on the knuckles for the bright but careless student who was secure in his understanding of the experiment and the calculation involved, but it most decidedly is not appropriate for the ordinary introductory student becoming acquainted with the computer for the first time. That part of the program now reads:

```
[152] RETURN←RETURN, (I26)+1
[153] ' |6|  MOLAR VOLUME OF OXYGEN (IN LITERS)'
```

```
      CHEMLAB1

PLEASE ENTER YOUR NAME, FIRST NAME FOLLOWED BY LAST.
KEVIN KELLEY
OK KEVIN LET'S GET ON WITH IT!

       DO YOU REQUIRE ASSISTANCE ? (TYPE YES OR NO)
⎕:
      NO

       TRIAL ? (TYPE 1 OR 2)
⎕:
      1
       SAMPLE DESIGNATION ? (IF PURE KCLO3, TYPE PURE)
⎕:
      A
  PART 1:  ENTER THE FIRST SIX PIECES OF DATA; SEPARATE ENTRIES WITH COMMAS.
        |1|     |2|     |3|    |4|    |5|    |6|
⎕:
      35.2347, 34.9161, 31.7249, 24.2, 74.2, 250

      DATA OUTSIDE RANGE OF PRESSURE CORRECTION TABLE; 710 ≤ P ≤ 769MM. HG
|5|   BAROMETRIC READING; UNCORRECTED AT ROOM TEMPERATURE.
⎕:
      742
      HAVE YOU MADE ANY ERRORS IN ENTERING YOUR DATA ?
⎕:
      NO

       YOUR EXPERIMENTAL VOLUME IS IN VERY GOOD AGREEMENT WITH THE THEORETICAL VOLUME(UN-
      CORRECTED)
  FOR YOUR SAMPLE AND YOUR EXPERIMENTAL CONDITIONS. PROCEED WITH YOUR CALCULATIONS.
     WHEN YOU NEED ME AGAIN, JUST CALL (CHEMLAB1)!
```

Fig. 7. Vector data entry.

```
[154]→((TEMP≥10)∧((TEMP←⎕≤50))/AROUND4
[155] 'MOLAR VOLUME IN LITERS, PLEASE!
[156]→RETURN[⍴RETURN]
[157] AROUND4:DATA[I;SWITCH;14]←TEMP
```

It can be seen that if the test in step 154 is not met successfully, a more appropriate message is given (step 155), and the program returns to the question (step 153) for a repeat.

It is important to place checks for decimal points or proper dimensions in a program, with responses that will permit a student to find an acceptable answer before bouncing him from the program, or introducing an error signal. For instance, an early version of the CHEMLAB program had the following sequence:

```
[62] '|4| TEMPERATURE OF THE OXYGEN [CENTIGRADE]:'
[63] DATA [I;SWITCH;5]←⌊TEMP←⎕
[64]→(0.5>TEMP-DATA[I;SWITCH;5])/AROUND
[65] DATA[I;SWITCH;5]←DATA[I;SWITCH;5]+1
[66] AROUND:(⍳26)+FUDGY
[67] RETURN←RETURN,(⍳26)+1
[68] '|5| BAROMETRIC READING; UNCORRECTED AT'; DATA[I;
     SWITCH;5]; ' °CENTIGRADE'
```

In step 62, the student is requested to respond to a question concerning gas temperature. His numerical answer, indicated by the open box at the extreme right-hand end of step 63, is stored in a location called "TEMP". This value is shorn of any decimal part (by the operator "⌊") and the resulting integer is stored in an array location named "DATA[I; SWITCH;5]". Step 64 checks to see if the difference between "TEMP" (which has no fractional part) and "DATA[I;SWITCH;5]" (an integer) is less than 0.5. If so, the rounded value is properly stored, and the program is branched to that step labelled "AROUND" (step 66), which in turn sends the program to step 68--a new question. However, if the value found in step 64 is not less than 0.5, the program does not branch to AROUND, but falls through to step 65, which increases the value stored in DATA[I;SWITCH;5] by 1. These four steps in effect round off the temperature values entered to the nearest degree. This temperature value, stored in DATA[I;SWITCH;5], will be used later as an index to pick out a vapor pressure correction from a table stored in the workspace (a later version of the program performs this four-line operation in one part of one line). This value of the vapor pressure of water at the temperature of the experiment is then used in a calculation of a computer-derived result which is used as the basis for comparison with the student-calculated result.

The programming problem and solution outlined above ignores two possible alternative responses: the temperature measured may fall outside either extreme of the temperature correction table, or the student may make a typing or understanding error in entering the datum so that the number entered is not a temperature value at all. Since this is the kind of error such a program should recognize, a checking sub-routine has been introduced:

```
[52] RETURN←RETURN, (I26)+1
[53] '|4| TEMPERATURE OF THE OXYGEN [CENTIGRADE]:'
[54]→((TEMP≥16)∧((TEMP←⎕)≤33))/AROUND
[55] 'DATA OUTSIDE RANGE OF TEMPERATURE CORRECTION TABLE;
     16≤T≤33° C'
[56]→RETURN[⍴RETURN]
[57] AROUND:DATA[I;SWITCH;5]←TEMP
[58]→(I26)+FUDGY
[59] RETURN←RETURN, (I26)+1
[60] '|5| BAROMETRIC READING; UNCORRECTED AT ROOM
     TEMPERATURE.'
```

In this sequence, step 53 asks for the observed temperature, Step 54 tests the entered value to see if it falls within the range of 16-35° C. If this requirement is met, the program banches to the next "AROUND", which is step 57, and the temperature value is stored in DATA[I;SWITCH;5]. Should this requirement not be met, the program proceeds to the next line, which is an error message:

DATA OUTSIDE RANGE OF TEMPERATURE CORRECTION TABLE;
16≤T≤33° C

After this print-out, the next step, 56, returns the program to step 52, which begins the question and answer sequence again. Checks of this type have been introduced at many points in the program.

Another major feature of this program is the inclusion of a routine which permits a student to examine his input data, identify a mistake, and correct it without having to reenter all data (see III. C. 2). The steps which accomplish this are found in lines 72-82 of the CHEMLAB1 listing. Since this question-retry routine is one of the more opaque segments of this program, a detailed analysis of the steps follows. Some earlier steps are included in the analysis, since they set indicators necessary for the branching routines.

Line 37: This line is the beginning of Part 1.

Line 39: FUDGY is an index to control branching. It is initially set to 1 so that the operation →(I26)+FUDGY will yield a branch to the succeeding line.

Later on, FUDGY will receive values which will cause branching to predetermined locations within the function.

Line 40: RETURN is a vector composed of the statement numbers associated with data input statements. It is constructed by catenating together the statement numbers of these lines as the student makes his first pass through the input statements. Should the student require an updating of information already entered, a branch can be made back to the appropriate line by way of these statement numbers. A statement of this kind is used before every data entry.

Line 43: Line 43 is a variable branch statement. When FUDGY has a value of 1, the branch is to the following line. When FUDGY has some other value (calculated in line 81), the branch is to that statement specified by I26 (the statement being executed) plus the value of FUDGY. This statement appears after each data entry.

Line 77: RETURN is the vector of statement numbers; its last element is the statement number associated with the beginning of the Question Retry Area. GOTO is a vector composed of the statement numbers associated with only those areas the student wishes to retry. GOTO's last element is the same as RETURN's last element.

Line 78: GATE is a matrix composed of question entry points matched with corresponding exit points. Example: Line 49 is the entry point for Question #3, and Line 51 is its exit point.

Line 79: This line is the Question Retry Area termination check. It checks to see if GOTO has fewer than two elements; this should occur only after the last student retry request has been processed.

Line 80: This line truncates from GOTO its first element; this element is the statement number associated with the question that has just been processed. On the first pass, this element is set to zero and truncated.

Line 81: This line calculates the value of FUDGY necessary to cause a branch back to Line 79 which is the Question Retry Area termination check. Note that this branch-back will occur from the termination point of the question being retried.

Line 82: This is a branch to the first element of GOTO, which is always a statement number and always a question entry point.

The relatively elaborate data entry and correction features just described make it easy for students new to the computer to use it without frustrating delay. These kinds of features are especially important in the programming of functions to be used by a large number of students within a specified, limited period of time.

## V. CONCLUSION

The CHEMLAB program has been described in some detail because in a number of ways it is our most successful program to date, and it represents a kind of use which we have not seen described elsewhere. We are encouraged by its reception and are convinced that this is, indeed, a useful way to use the computer in the laboratory. The program was initially written in the summer of 1970 and was first used by a small class in that summer session of general chemistry. The experience with the summer group led to fairly extensive modifications and expansion, and the revised program was used by over 300 students in the fall semester of 1970. Ready access to primary data for a large class permitted us to examine closely the results of the experiment, to set reasonable limits of acceptability on the experimental results, and to identify some of the recurrent trouble spots in the experiment. We were then better able to focus on these points in the pre-lab discussion. Further use by large classes in the spring and fall of 1971 permitted us to evaluate some of the programming features described in Section IV.

Although a few students regarded the use of the computer as an extra chore, the general response of the students can be described only as enthusiastic and excited. Typical student responses to an additudinal survey are shown in Fig. 8. A large majority of the students felt that the computer experience was both useful and exciting. Less than 10% of the class indicated a strongly negative reaction. It is interesting that, in this age of "Do your own thing," over 60% of the students have suggested that the use of CHEMLAB should be required rather than optional. Indeed, the most frequently stated student suggestion was that we should extend this kind of program to other lab experiments.

Despite the fact that use of the computer by the entire student group during the laboratory period occasionally caused some additional confusion and extra work for the lab instructors, their reaction to CHEMLAB has been quite favorable. The program has certainly been helpful and time-saving in evaluating student lab performance and grading reports. It has encouraged the student to learn to do the necessary calculations and data manipulations correctly, since calculation errors are immediately evident to both student and instructor.

| | Questions | Percent of Responses | | |
|---|---|---|---|---|
| | | Definitely | Somewhat | No |
| 1) | Was the computer program helpful to you in evaluating your experimental data? | 73 | 24 | 3 |
| 2) | Did you find using the computer was difficult and confusing? | 4 | 19 | 77 |
| 3) | Do you feel that using the computer saved you time by pointing out experimental or calculation errors? | 64 | 25 | 11 |
| 4) | Do you feel that using the computer was just an additional "chore", and contributed very little to the experiment? | 3 | 6 | 91 |
| 5) | Did your computer program results encourage you to think about the principles and procedures of the experiment? | 21 | 59 | 20 |
| 6) | Did your computer results help you in your lab report discussion? | 48 | 34 | 18 |
| 7) | Was the experience of using the computer interesting and/or exciting for you? | 78 | 19 | 3 |
| 8) | Are you interested in learning more about APL and other computer languages? | 75 | 16 | 9 |
| 9) | Did the use of the computer take more time than it was worth? | 5 | 16 | 79 |
| 10) | Should use of the computer for this experiment be optional rather than mandatory? | 29 | 7 | 64 |
| 11) | Is this the first time you have used a computer? | 75 | | 25 |

Fig. 8. Attitudinal survey of students in introductory chemistry concerning CHEMLAB.

It would seem, then, that the first three specific objectives for CHEMLAB (as listed in Section III. B) have been well met. Accomplishment of the fourth objective is less conclusively established. There has been some improvement in both lab reports and student questions. Yet, both student attitude surveys (Fig. 8, questions 5 and 6) and lab instructors' comments indicate we have had only limited success in encouraging the student to think critically about the experiment. We are convinced, however, that we have made a step in the right direction, and that this particular kind of in-lab computer use can be helpful in accomplishing this objective.

Adjunct use of "canned" programs like CHEMLAB can be criticized as being "black-box" exercises. Although the student is aware of what the program does, he is not gaining experience in programming or even in formulating a suitable function or algorithm to solve his problem or treat his data. Nevertheless, we feel this type of program does have real value, especially at the introductory level. In addition to its immediate value as an aid in data treatment, problem solving, or conceptual understanding, it can serve to help students overcome a fear of the computer and make them aware of some of its potential uses. This, in turn, encourages individual effort in programming and computer use. For those who do become interested, sufficient courses and guidance are available. The canned programs can serve as models, since all of our APL teaching functions are stored in public libraries and are available from the terminal to any user. As our CAI effort expands, and more students become exposed early to interactive use of the computer in a number of courses, the transition from canned programs to student-written programs should take place more easily.

Although the cost of introducing CAI and developing programs for class use has been high, these initial efforts served as a focus for the introduction of a number of students and faculty in Chemistry to the computer. With such programs in hand, further developments have been much easier to implement.

Figure 9 briefly describes many of the APL programs which have been used by students and faculty in our department. We have no doubt that the use of the computer in this way has added to our effectiveness in teaching chemistry--which is, after all, our ultimate aim. The development of random access slide projector capability, CRT capability, and other peripherals for the APL system should serve to strengthen its position as a useful and flexible interactive system for computer-aided instruction.

FUNCTIONS CONTAINED IN APL PUBLIC LIBRARY 18

| FUNCTION NAME | DESCRIPTION | WORKSPACE |
|---|---|---|
| ANAL | COMPOSITION ANALYSIS AID | 18 ANALYT |
| BENAE | CALCULATES ACTIVATION ENERGY | 18 CHEM2 |
| BENK | CALCULATES ENTHALPY FROM TEMP. DEPENDENCE | 18 CHEM2 |
| CHEMLAB1 | LABORATORY CAI IN DECOMPOSITION | 18 CHEMLAB2 |
| CRINGE | CONVERSATIONAL GRADE TAB. PROGRAM | 18 GRADER |
| ENTIRE | CONTENTS OF CATALOG18 | 18 CATALOG18 |
| GC | GAS CHROMATOGRAPHY CALC. | 18 ANALYT |
| GOLD | STUDY OF CELLULAR AUTOMATA | 18 PATTERN |
| INSTRUCTOR | ANALYZES RESULTS FROM CHEMLAB1 | 18 LABINSTRUCT |
| KSP | SIMULATES SOLUBILITY OF GIVEN SALT | 18 CHEM3 |
| PICTURE | PICTURE OF ONE-SIDED POTENTIAL WELL | 18 QUANTUMINFO |
| PSIFNS | GROUP OF WAVEFUNCTIONS FOR ONE-SIDED WELL | 18 QUANTUM1 |
| RATE1C | RATE CONSTANT FROM CONCENTRATION DATA | 18 CHEM1 |
| RATE2C | RATE CONSTANT FROM CONCENTRATION DATA | 18 CHEM1 |
| RATE1OD | RATE CONSTANT FROM ABSORBANCE DATA | 18 CHEM1 |
| RATE1ODN | RATE CONSTANT FROM ABSORBANCE DATA | 18 CHEM1 |
| SEQU | CAI: KINETIC REACTIONS | 18 REACT |
| SEQU3 | CONTINUATION OF REACTX | 18 REACT3 |
| SEQU2 | CAI: COMPLEX KINETIC REACTIONS | 18 REACT2 |
| SEQUX | CONTINUATION OF REACT2 | 18 REACTX |
| TITRATE | SIMULATES TITRATION OF ACID AND BASE | 18 CHEM3 |
| TRY | EIGENVALUES OF A ONE-SIDED POTENTIAL WELL | 18 QUANTUM1 |
| UTILITIES | GROUP OF USEFUL??? UTILITY FUNCTIONS | 18 QUANTUM1 |
| ENT | CALCULATES ENTROPY OF GAS FROM SPECT. DATA | 18 ENTROPY |
| IONIZATION1 | GRAPHIC ILLUST. OF MULTISTEP EQUILIBRIA | 18 ION |
| COMPLEXION | GRAPHIC ILLUST. OF MULTISTEP EQUILIBRIA | 18 ION |
| POLYPROTIC | GRAPHIC ILLUST. OF MULTISTEP EQUILIBRIA | 18 ION |

Fig. 9. Functions contained in APL Chemistry Library, SUNY-Binghamton.

## ACKNOWLEDGMENTS

We are particularly indebted to Mr. James Higgins, academic manager of the SUNY-Binghamton computer center, for his encouragement and support of the development of our computer applications in Chemistry. We were very fortunate in having available to us the programming assistance and expertise of Mr. Kevin Kelley and Ms. Anne Kellerman. This manuscript would never have been delivered without the patient and understanding excellence of Marlene, for which we are grateful.

# APPENDIX

## CHEMLAB1 LISTING

```
      ∇ CHEMLAB1
[1]     ''
[2]     ''
[3]     L←1+K←1+J←1+I←1+H←1+G←1+F←1+E←1+D←1+C←1+B←1+A←1+PURE←1
[4]     YES←1+NO←0
[5]     FIND
[6]     →(V̲Δ̲<1)/0                         (FIND is an auxilliary function which matches a student name
[7]     →(DATA[V̲Δ̲;1;15]≠¯1)/MATCH         entry, against a previously defined list.)
[8]     DATA[V̲Δ̲;1;15]←1
[9]     ''
[10]    '        DO YOU REQUIRE ASSISTANCE ? (TYPE YES OR NO)'
[11]    →(~⎕)/MATCH
[12]    ''
[13]    '        WHEN YOU ENTER DATA, ALL WEIGHTS SHOULD BE IN GRAMS, ALL VOLUMES IN MILLI
        LITERS, AND'
[14]    ' ALL PRESSURES IN MILLIMETERS[OF HG], U̲N̲L̲E̲S̲S̲ SPECIFIED OTHERWISE.  '
[15]    ''
[16]    '        D̲O̲ N̲O̲T̲ INCLUDE UNITS WHEN ENTERING DATA.'
[17]    ''
[18]    '        DATA IS ENTERED BY TYPING IN THE DESIRED CHARACTERS AND PRESSING THE ''RE
        TURN'' KEY.'
[19]    '  IF A MISTAKE IS MADE IN TYPING, DO N̲O̲T̲ PRESS THE ''RETURN'' KEY; USE THE ''BAC
        KSPACE'' KEY TO'
[20]    '  BACKSPACE THE PRINTING ELEMENT UNTIL IT IS POSITIONED DIRECTLY OVER THE FIRST
        (ON LEFT) FAULTY'
[21]    '  CHARACTER.  DEPRESS THE ''ATTN'' KEY AND REPLACE THE FAULTY CHARACTER(S) BY TY
        PING IN THE CORRECT'
[22]    '  CHARACTER(S).  RETYPE A̲L̲L̲ CHARACTERS TO THE RIGHT OF THE CHARACTER(S) REPLACED
        .  PRESS THE ''RETURN'' KEY.'
[23]    ''
[24]    '        IF A MISTAKE HAS ALREADY BEEN ENTERED BY PRESSING ''RETURN'', CONTINUE ON
         UNTIL I ASK IF ANY'
[25]    '  MISTAKES HAVE BEEN MADE.'
[26]  MATCH:''
[27]    '        TRIAL ? (TYPE 1 OR 2)'
```

```
[28]  SWITCH←⎕
[29]  →(DATA[V∆;SWITCH;1]≠14)/DRACULA
[30]  '        SAMPLE DESIGNATION ? (IF PURE KCLO3, TYPE PURE)'      ⎤ (Branches student to
[31]  DATA[V∆;SWITCH;1]←NUMBAH←⎕                                      ├ appropriate Part of
[32] DRACULA:→((⍳4)∊DATA[I←V∆;SWITCH;15])/FAR,FURTHER,FINALE,FAREST ⎦ program.)
[33] FAREST:'       MY RECORDS INDICATE YOU HAVE ALREADY ENTERED VALID INFORMATION FOR
      THE SAMPLE'
[34]  '  TYPE AND TRIAL CHOSEN.  THIS INFORMATION CANNOT BE ALTERED.  IF THERE IS A VAL
      ID REASON TO CHANGE'
[35]  '  THIS DATA, INDICATE IT IN YOUR LABORATORY REPORT.'
[36]  →0
[37] FAR:'  PART 1:  ENTER THE FOLLOWING PIECES OF DATA:'
[38]  ''
[39]  FUDGY←PTR←1                                                        ⎤
[40]  RETURN←(⌶26)+1                                                     │
[41]  '|1|   WEIGHT OF TUBE AND CONTENTS BEFORE HEATING.'                │
[42]  DATA[I;SWITCH;2]←⎕                                                 │
[43]  →(⌶26)+FUDGY                                                       │
[44]  RETURN←RETURN,(⌶26)+1                                              │
[45]  '|2|   WEIGHT OF TUBE AND RESIDUE AFTER HEATING.'                  │
[46]  DATA[I;SWITCH;3]←⎕                                                 │
[47]  →(⌶26)+FUDGY                                                       │
[48]  RETURN←RETURN,(⌶26)+1                                              ├ (DATA entry
[49]  '|3|   WEIGHT OF EMPTY TUBE.'                                      │  for Part I)
[50]  DATA[I;SWITCH;4]←⎕                                                 │
[51]  →(⌶26)+FUDGY                                                       │
[52]  RETURN←RETURN,(⌶26)+1                                              │
[53]  '|4|   TEMPERATURE OF THE OXYGEN [CENTIGRADE]:'                    │
[54]  →((TEMP≥16)∧((TEMP←⎕)≤33))/AROUND                                  │
[55]  '      DATA OUTSIDE RANGE OF TEMPERATURE CORRECTION TABLE; 16≤T≤33 °C'
[56]  →RETURN[⍴RETURN]                                                   │
[57] AROUND:DATA[I;SWITCH;5]←TEMP                                        │
[58]  →(⌶26)+FUDGY                                                       │
[59]  RETURN←RETURN,(⌶26)+1                                              │
[60]  '|5|   BAROMETRIC READING; UNCORRECTED AT ROOM TEMPERATURE.'       │
```

```
[61]  →((TEMP≥710)∧((TEMP←⎕)≤769))/AROUND2
[62]  '      DATA OUTSIDE RANGE OF PRESSURE CORRECTION TABLE; 710≤P≤769 MM. HG'
[63]  →RETURN[⍴RETURN]
[64] AROUND2:DATA[I;SWITCH;8]←TEMP
[65]  →(⌶26)+FUDGY
[66]  RETURN←RETURN,(⌶26)+1                                     (DATA entry for Part I)
[67]  '|6|    VOLUME (UNCORRECTED) OF OXYGEN COLLECTED.'
[68]  DATA[I;SWITCH;7]←⎕
[69]  →(⌶26)+FUDGY
[70]  RETURN←RETURN,LAST←(⌶26)+1
[71]  ''
[72]  '      HAVE YOU MADE ANY ERRORS IN ENTERING YOUR DATA ?'
[73]  →(~⎕)/(PTR=⍳3)/GOTCHA,GOAL,GOLLY
[74]  ''
[75]  '      ENTER THE NUMBER OR NUMBERS ( FOUND IN THE | | ) OF THE INPUT STATEMENTS A
      SSOCIATED WITH'
[76]  ' THE DATA YOU WISH TO CORRECT.  IF MORE THAN ONE NUMBER IS TO BE ENTERED, SEPAR
      ATE WITH BLANKS.'
[77]  GOTO←0,(((⍳(DQ←⍴RETURN)-1)∊⎕),1)/RETURN
[78]  GATE←(2,DQ)⍴RETURN,1⌽(RETURN-2)
[79]  →((⍴GOTO)≤2)/RETURN[DQ]
[80]  GOTO←1↓GOTO
[81]  FUDGY←(⌶26)-GATE[2;(GATE[1;]∊GOTO[1])/⍳DQ]+2
[82]  →GOTO[1]
[83] GOTCHA:PART←((20×⌊0.5+DATA[I;SWITCH;8]÷20)-700)÷20
[84]  PART←DATA[I;SWITCH;8]-TCTABLE[⌊0.5+DATA[I;SWITCH;5]-15;PART]+    (Computer calc. of
      0.03                                                             theoretical volume)
[85]  PART←PART-VPTABLE[⌊0.5+DATA[I;SWITCH;5]-15]
[86]  DATA[I;SWITCH;6]←(PERCENT[NUMBAH]×(DATA[I;SWITCH;2]-DATA[I;SWITCH;4])×25528020×(
      DATA[I;SWITCH;5]+273.16))
[87]  DATA[I;SWITCH;6]←DATA[I;SWITCH;6]÷33475.758×PART
[88]  FUDGE←|(DATA[I;SWITCH;6]-DATA[I;SWITCH;7])÷DATA[I;SWITCH;6]    (Test of agreement
[89]  ''                                                               between experimental
[90]  →((FUDGE<0.1),((0.1≤FUDGE)∧(FUDGE≤0.2)),((0.2≤FUDGE)∧(FUDGE≤    and theoretical
      0.3)),(FUDGE>0.3))/VG,RG,PG,UA                                   values)
```

```
[91] VG:'        YOUR EXPERIMENTAL VOLUME IS IN VERY GOOD AGREEMENT WITH THE THEORETICAL
       VOLUME(UNCORRECTED)'
[92]   '  FOR YOUR SAMPLE AND YOUR EXPERIMENTAL CONDITIONS. PROCEED WITH YOUR CALCULATIO
       NS.'
[93]   →SMUDGE
[94] RG:'        YOUR EXPERIMENTAL VOLUME IS IN REASONABLY GOOD AGREEMENT WITH THE THEOR
       ETICAL VOLUME'
[95]   '  (UNCORRECTED)FOR YOUR SAMPLE AND YOUR EXPERIMENTAL CONDITIONS. PROCEED WITH YO
       UR CALCULATIONS'
[96]   '  ON THIS SAMPLE. YOU MAY WISH TO REPEAT THE EXPERIMENT(TRIAL 2) TO OBTAIN MORE
       ACCURATE RESULTS.'
[97]   →SMUDGE
[98] PG:'        YOUR EXPERIMENTAL VOLUME, ALTHOUGH ACCEPTABLE IS IN FAIRLY POOR AGREEME
       NT WITH THE'
[99]   '  THEORETICAL VOLUME(UNCORRECTED)FOR YOUR SAMPLE AND YOUR CONDITIONS. YOU MAY'
[100]  '  PROCEED WITH YOUR CALCULATIONS ON THIS SAMPLE IF YOU WISH, BUT YOU ARE ALSO UR
       GED'
[101]  '  TO REPEAT THE EXPERIMENT (TRIAL 2) TO OBTAIN BETTER RESULTS.'
[102]  →HORSE
[103]UA:'        YOUR EXPERIMENTAL VOLUME IS UNACCEPTABLE(ERROR>30PERCENT).PLEASE REPEAT
        THE'
[104]  '  EXPERIMENT.  YOU WILL BE ALLOWED TO ENTER YOUR NEW DATA UNDER THE SAME TRIAL N
       UMBER.'
[105]HORSE:→((DATA[I;SWITCH;7]>DATA[I;SWITCH;6]),DATA[I;SWITCH;7]<DATA[I;SWITCH;
       6])/YHIGH,YLOW
[106]YHIGH:'        YOUR MEASURED VOLUME IS TOO HIGH.  CONSIDER THE FOLLOWING POSSIBLE S
       OURCES OF ERROR:'
[107]  SPACE 12;'DID YOU ENTER WEIGHTS CORRECTLY?'
[108]  SPACE 12;'DID YOU ALLOW THE SYSTEM TO COOL TO NEAR ROOM TEMPERATURE?'
[109]  →(FUDGE>0.3)/0
[110]  →SMUDGE
[111]YLOW:'        YOUR MEASURED VOLUME IS TOO LOW. CONSIDER THE FOLLOWING POSSIBLE SOUR
       CES OF ERROR:'
[112]  SPACE 12;'DID YOU ENTER WEIGHTS CORRECTLY?'
[113]  SPACE 12;'DID YOU COMPLETELY DECOMPOSE THE KCLO3 IN YOUR SAMPLE?  (DID YOU HEAT L
       ONG ENOUGH)'
[114]  SPACE 12;'DID YOU HAVE AN AIRTIGHT SYSTEM (NO LEAKS!)?'
```

```
[115] →(FUDGE>0.3)/0
[116]SMUDGE:WT←PERCENT[NUMBAH]×(DATA[I;SWITCH;2]-DATA[I;SWITCH;4])×(48÷     (Test of agree-
      122.55)                                                              ment between
[117] FUDD←|(WT-(DATA[I;SWITCH;2]-DATA[I;SWITCH;3]))÷WT                     exptl. & theoret.
[118] →((FUDD>0.25),((0.1≤FUDD)∧(FUDD≤0.25)))/UWT,PWT                       weight loss)
[119] DATA[I;SWITCH;15]←2
[120] '      WHEN YOU NEED ME AGAIN, JUST CALL (CHEMLAB1)!  '
[121] →0
[122]UWT:'HOWEVER, YOUR EXPERIMENTAL WEIGHT LOSS IS SERIOUSLY IN ERROR AND UNACCEPTABLE
      (>25PERCENT).'
[123] 'DO NOT PROCEED WITH CALCULATIONS.  IF WEIGHTS ENTERED INCORRECTLY, CALL CHEMLAB1
       AGAIN.  '
[124] →0
[125]PWT:'YOUR EXPERIMENTAL WEIGHT LOSS IS IN CONSIDERABLE ERROR (10-25 PERCENT). CHECK
       YOUR WEIGHTS.'
[126] DATA[I;SWITCH;15]←2
[127] →0
[128]FURTHER:' PART 2:   ENTER THE FOLLOWING PIECES OF DATA:'
[129] NUMBAH←DATA[I;SWITCH;1]
[130] FUDGY←1
[131] RETURN←(ɪ26)+1
[132] '|1|   WEIGHT OF OXYGEN COLLECTED.'
[133] DATA[I;SWITCH;9]←⎕
[134] →(ɪ26)+FUDGY
[135] RETURN←RETURN,(ɪ26)+1
[136] '|2|   TEMPERATURE, ABSOLUTE.'
[137] DATA[I;SWITCH;10]←⎕
[138] →(ɪ26)+FUDGY
[139] RETURN←RETURN,(ɪ26)+1
[140] '|3|   BAROMETRIC PRESSURE, CORRECTED'
[141] SPACE 13;'( TOTAL GAS PRESSURE )'
[142] DATA[I;SWITCH;11]←⎕
[143] →(ɪ26)+FUDGY
[144] RETURN←RETURN,(ɪ26)+1
[145] '|4|   PRESSURE DUE TO OXYGEN'
[146] DATA[I;SWITCH;12]←⎕
[147] →(ɪ26)+FUDGY
```

```
[148] RETURN←RETURN,(ɪ26)+1
[149] '|5|    VOLUME OF OXYGEN AT STP'
[150] DATA[I;SWITCH;13]←⎕
[151] →(ɪ26)+FUDGY
[152] RETURN←RETURN,(ɪ26)+1
[153] '|6|    MOLAR VOLUME OF OXYGEN ( IN LITERS )'
[154] →((TEMP≥10)∧((TEMP←⎕)≤50))/AROUND4
[155] 'MOLAR VOLUME IN LITERS, PLEASE! '
[156] →RETURN[ρRETURN]
[157]AROUND4:DATA[I;SWITCH;14]←TEMP
[158] →(ɪ26)+FUDGY
[159] RETURN←RETURN,(1+ɪ26),LAST
[160] PTR←2
[161] →LAST
[162]GOAL:TEMP←(DATA[I;SWITCH;2]-DATA[I;SWITCH;3]),(273.2+DATA[I;SWITCH;5])  ⎤
[163] TEMP←TEMP,,(DATA[I;SWITCH;8]-TCTABLE[⌊0.5+DATA[I;SWITCH;5]-15;((20×⌊   ⎥ (Calc. from
      0.5+DATA[I;SWITCH;8]÷20)-700)÷20]+0.03)                                  ⎥ student pri-
[164] TEMP←TEMP,TEMP[3]-VPTABLE[⌊0.5+DATA[I;SWITCH;5]-15]                     ⎥ mary data)
[165] TEMP←TEMP,(DATA[I;SWITCH;7]×273.16×TEMP[4])÷(TEMP[2]×760)              ⎥
[166] TEMP←TEMP,(TEMP[5]÷(TEMP[1]÷31.9988))÷1000                              ⎦
[167] TEMP2←(DATA[I;SWITCH;2]-DATA[I;SWITCH;4])×PERCENT[NUMBAH]×(  ⎤ (Calc. of theoretical
      47.9982÷122.5532)                                              ⎥ values)
[168] TEMP2←TEMP2,TEMP[2 3 4],((TEMP2÷31.9988)×22.4×1000)           ⎥
[169] TEMP2←TEMP2,(TEMP2[5]÷(TEMP2[1]÷31.9988))÷1000                ⎦
[170] ERROR←|(TEMP2-,DATA[I;SWITCH; 9 10 11 12 13 14])÷TEMP2
[171] ERROR2←(|(TEMP-,DATA[I;SWITCH; 9 10 11 12 13 14])÷TEMP)>0.05
[172] ERROR← 4 6 ρ((ERROR>0.05)∧(ERROR<0.1)),((ERROR≥0.1)∧(ERROR<
      0.2)),((ERROR≥0.2)∧(ERROR≤0.4)),ERROR>0.4
[173] SPACE 58;'|¯¯¯¯¯¯¯¯¯¯¯¯¯MACHINE COMPUTED USING YOUR DATA.'
[174] SPACE 42;'YOUR';SPACE 12;'|'
[175] SPACE 42;'DATA';SPACE 12;'↓';SPACE 10;'↓¯¯MACHINE COMPUTED THEORETICAL VALUES.'
[176] CHECKR 1
[177] 11 4 DFT TEMP[1] AND TEMP2[1] VS,DATA[I;SWITCH;9]
[178] '     WEIGHT OF OXYGEN.                ';Z[1;],CHECK
[179] CHECKR 2
[180] 11 1 DFT TEMP[2 3 4] AND TEMP2[2 3 4] VS,DATA[I;SWITCH; 10 11 12]
[181] '     ABSOLUTE TEMPERATURE.        ';Z[1;],CHECK
```

```
[182] CHECKR 3
[183] '      CORRECTED BAROMETRIC READING.';Z[2;],CHECK
[184] CHECKR 4
[185] '      PRESSURE DUE TO OXYGEN.       ';Z[3;],CHECK
[186] CHECKR 5
[187] 11 0 DFT TEMP[5] AND TEMP2[5] VS,DATA[I;SWITCH;13]
[188] '      VOLUME OF O2 AT STP.          ';Z[1;],CHECK
[189] CHECKR 6
[190] 11 1 DFT TEMP[6] AND TEMP2[6] VS,DATA[I;SWITCH;14]
[191] '      MOLAR VOLUME OF O2.            ';Z[1;],CHECK
[192] ''
[193] '        COMPARISON OF THESE COLUMNS ALLOWS YOU TO CHECK BOTH YOUR CALCULATIONS AN
      D'
[194] '  THE AGREEMENT BETWEEN YOUR EXPERIMENTAL DATA AND THE THEORETICAL VALUES FOR YO
      UR SAMPLE.'
[195] ''
[196] →((+/,ERROR)=0)/HOOP
[197] 'ERROR MESSAGE INTERPRETATION:'
[198] '          *   5 - 10 PERCENT ERROR.'
[199] '         **  10 - 20 PERCENT ERROR.'
[200] '        ***  20 - 40 PERCENT ERROR.'
[201] '          ';ARGG;'      >40 PERCENT ERROR.'
[202]HOOP:→((+/ERROR[4;])=0)/LOCKET
[203] '        YOUR DATA IS UNACCEPTABLE(ERROR>40 PERCENT); DID YOU ENTER DATA CORRECTLY
      ?'
[204] '  ARE YOUR CALCULATIONS CORRECT?'
[205] '        IF BOTH YOUR DATA AND CALCULATIONS ARE CORRECT, THEN CONSULT YOUR LAB INS
      TRUCTOR. IF'
[206] '  EITHER DATA ENTRY OR CALCULATIONS ARE INCORRECT, THEN YOU MAY CORRECT THESE AN
      D RE-'
[207] '  ENTER VALUES FOR THE SAME TRIAL NUMBER.'
[208] →0
[209]LOCKET:'        PLEASE HAND-IN THIS TABLE WITH YOUR FINAL LAB REPORT.'
[210] '  IF YOUR SAMPLE IS AN UNKNOWN, YOU MAY PROCEED TO PART 3, WHEN READY, BY TYPING
       CHEMLAB1 AGAIN.'
[211] DATA[I;SWITCH;15]←3
```

(Setting up table and error messages for Part II)

```
[212] →0
[213] ''
[214] ''
[215]FINALE:'  PART 3:  ENTER THE FOLLOWING DATA:'
[216] FUDGY←1
[217] RETURN←(ι26)+1
[218] '|1|   MOLES OF OXYGEN.'
[219] DATA[I;SWITCH;16]←⎕
[220] →(ι26)+FUDGY
[221] RETURN←RETURN,(ι26)+1
[222] '|2|   MOLES OF POTASSIUM CHLORATE DECOMPOSED.'
[223] DATA[I;SWITCH;17]←⎕
[224] →(ι26)+FUDGY
[225] RETURN←RETURN,(ι26)+1
[226] '|3|   WEIGHT OF POTASSIUM CHLORATE IN SAMPLE.'
[227] DATA[I;SWITCH;18]←⎕
[228] →(ι26)+FUDGY
[229] RETURN←RETURN,(ι26)+1
[230] '|4|   WEIGHT OF SAMPLE USED.'
[231] DATA[I;SWITCH;19]←⎕
[232] →(ι26)+FUDGY
[233] RETURN←RETURN,(ι26)+1
[234] '|5|   PERCENT OF POTASSIUM CHLORATE IN SAMPLE.'
[235] →((TEMP←⎕)>1)/AROUND3
[236] 'I WANT A TRUE PERCENT; NOT A FRACTION !'
[237] →RETURN[ρRETURN]
[238]AROUND3:DATA[I;SWITCH;20]←TEMP
[239] →(ι26)+FUDGY
[240] RETURN←RETURN,(1+ι26),LAST
[241] PTR←3
[242] →LAST
[243]GOLLY:MOLE←,(DATA[I;SWITCH;2]-DATA[I;SWITCH;3])÷31.9988  ┐
[244] MOLE←MOLE,(MOLE[1]×0.6667)                              │ (Calc. from student
[245] MOLE←MOLE,(MOLE[2]×122.552)                             ├ primary data)
[246] MOLE←MOLE,(DATA[I;SWITCH;2]-DATA[I;SWITCH;4])           │
[247] MOLE←MOLE,(MOLE[3]÷MOLE[4])×100                         ┘
```

```
[248] 11 5 DFT MOLE[1 2] VS,DATA[I;SWITCH; 16 17]
[249] ''
[250] SPACE 48;'YOUR';SPACE 11;'|¯¯¯MACHINE COMPUTED USING YOUR DATA.'
[251] SPACE 48;'DATA';SPACE 11;'↓'
[252] ''
[253] '  MOLES OF OXYGEN EVOLVED.';SPACE 16;Z[1;]
[254] '  MOLES OF KCLO3 DECOMPOSED.';SPACE 14;Z[2;]
[255] 11 3 DFT MOLE[3 4] VS,DATA[I;SWITCH; 18 19]
[256] '  WEIGHT OF KCLO3 IN SAMPLE.';SPACE 14;Z[1;]
[257] '  WEIGHT OF SAMPLE USED.';SPACE 18;Z[2;]
[258] 11 1 DFT MOLE[5] VS,DATA[I;SWITCH;20]
[259] '  PERCENT KCLO3 IN SAMPLE.';SPACE 14;Z[1;]
[260] ''
[261] ''
[262] NUMBAH←DATA[I;SWITCH;1]
[263] '  THEORETICAL PERCENT POTASSIUM CHLORATE FOR YOUR SAMPLE IS ';(PERCENT[NUMBAH])×
      100;' PERCENT.'
[264] DATA[I;SWITCH;15]←4
[265] '  YOU HAVE FINISHED COMPUTER TREATMENT OF DATA FOR THIS SAMPLE, AND MAY NOT ALTE
      R'
[266] ' ANY OF THE DATA.  PLEASE HAND IN THIS TABLE WITH YOUR REPORT.  IF YOU HAVE MADE
       ANY'
[267] ' ERRORS IN YOUR FINAL CALCULATIONS, CORRECT THEM AND INCLUDE THE CORRECTIONS IN '
[268] ' YOUR LAB REPORT.  YOUR REPORT SHOULD INCLUDE A DISCUSSION OF YOUR EXPERIMENTAL
      ERROR. '
[269] →0
```

(Tabular output for Part III) [annotation bracketing lines 248–261]

## REFERENCES


1. a. "Conference on Computers in Chemical Education and Research", July 19-23, 1971, Northern Illinois University. b. "Second Conference on Computers in the Undergraduate Curricula", June 23-25, 1971, Dartmouth College. c. "Conference on Computers in the Undergraduate Curriculum", June 16-18, 1970, University of Iowa. d. "Computers in Undergraduate Education: Mathematics, Physics, Statistics and Chemistry", December 8-9, 1967, University of Maryland. e. F. D. Tabbutt, "Computers in Chemical Education", Chem. & Eng. News, January, 19, 1970, p. 44.
2. a. K. Iverson, A Programming Language, Wiley, New York, 1962. b. K. Iverson, "The Use of APL in Teaching", International Business Machines, New York 1969.
3. E. J. Corey, W. Todd Wipke, Richard D. Cramer III, and W. Jeffrey Howe, J. Am. Chem. Soc., 94:2, 421 (1972), and references cited therein.
4. A. W. Leuhrmann, "Second Conference on Computers in the Undergraduate Curricula," June 23-25, 1971, Dartmouth College, p. 1.
5. R. C. Grandey, J. Chem. Educ., 48:12, 791 (1971).
6. a. J. A. Young, J. Chem. Educ., 47:11, 758 (1970). b. Anthony G. Oetinger, with collaboration of Sema Marks, Run, Computer, Run: "The Mythology of Educational Innovation", Harvard Univ. Press, Cambridge, Massachusetts, 1969.
7. a. B. E. Norcross, R. H. Cartmell, and G. Hall, "The Hindered Rotation of Amides", presented at 158th National Americal Chemical Society Meeting, New York, September 8, 1969. b. Thomas R. Dehner and Bruce E. Norcross, "The Use of APL in Computer-Assisted Instruction in Freshman Chemistry", presented at 158th National Americal Chemical Society Meeting, New York, September 8, 1969.
8. a. Computer Instruction Issue, J. Chem. Educ., 47:2 (1970). b. Symposium on the Status of Computer-Assisted Instruction in Chemistry, Joint Conference of the Chemical Institute of Canada with the Americal Chemical Society, Toronto, Canada, May 27, 1970. c. R. C. Atkinson and H. A. Wilson, "Computers in Chemical Education", Science, 162, 73 (1968).
9. a. K. Iverson and A. Falkoff, APL/360 Users Manual, International Business Machine, New York, 1968. b. P. Berry, APL/360 Primer, International Business Machine, New York, 1970. c. Sandra Pakin, APL/360 Reference Manual, Science Research Associates, Inc., Chicago, Illinois, 1968. d. L. Gilman and A. J. Rose, APL/360, An Interactive Approach, Wiley, New York, 1970.

10. a. J. Higgins, ed., The March on Armonk, APL Users Conference, State University of New York at Binghamton, July 11-12, 1969.
b. SHARE XXXVI Conference, Los Angeles, March, 1971; SHARE XXXVII Conference, New York City, August, 1971.

11. D. S. Olson, S. R. Carroll, L. A. King, and J. F. Altenburg, "Symposium on the Status of Computer-Assisted Instruction in Chemistry", Joint Conference of the Chemical Institute of Canada with the American Chemical Society, Toronto, Canada, May 27, 1970.

Chapter 2

# COMPUTER-BASED TEACHING OF ORGANIC CHEMISTRY

Stanley G. Smith and James R. Ghesquiere

The Roger Adams Laboratory
Department of Chemistry
University of Illinois
Urbana, Illinois



## I. INTRODUCTION

The rapidly increasing importance of chemistry to the development of many areas of scientific inquiry has resulted in increased demand for high quality training in chemistry for students with diverse backgrounds and educational objectives. A tutorial procedure [1] in which each student has the complete attention of a knowledgeable, patient professor who is responsive to his learning rate and goals could meet this need, but this is simply not possible because of the many students who must be helped.

Computer-based eductional systems make it possible to approximate closely a one-to-one correspondence between professor and student. Such computer-based teaching systems [2] can respond to the individual needs of a student, patiently providing extra help where needed, moving the student ahead where appropriate, and providing examples of interest--that is, tailoring the course of study to the individual. The impersonal character of the machine is compensated for by its responsiveness to the student and by the private character of the discourse.

Although complex and comprehensive computer-based teaching programs can be prepared, many aspects of classroom teaching are best done by an instructor. Unfortunately, in many classroom situations the teacher spends time doing things best performed by a machine, often leaving unattended tasks for which humans are uniquely qualified. For example, in an introductory organic chemistry course topics such as nomenclature, simple numerical problems, functional group interconversions, and many reaction sequences are easily and effectively handled with the one-to-one tutorial and problem-solving capabilities of computer systems currently being developed [3-5]. Releasing the instructor from these tasks would permit class time to be spent on more subtle points of theory, reaction mechanisms, complex synthesis, and topics of current interest. In addition, preliminary data suggest that discussion sessions have increased value to the students when they have been preceded by computer-based instruction designed to provide needed background material [6].

The ability of the computer system to provide individualized instruction on a specific topic at any time makes it possible for a student whose preparation in some areas is inadequate to review a portion of a previous course at a time convenient to him rather than, for example, repeating an entire course. Lesson material prepared on the PLATO III system has often been used for this purpose. For example, a program on nuclear magnetic resonance has been used as an introduction in elementary courses and as a review for graduate students.

Computer-based instruction has an interesting and productive tutorial application for students before they undertake actual laboratory work. The student interacts with the programs, on an individual basis and at his own rate, until he has acquired an understanding of the particular theory, objectives, and experimental procedures. Although our work in this area has just begun, it is anticipated that the computer-based development of the theory and simulations of important aspects of the experiment combined with actual laboratory work will lead to better understanding of the experiment and more efficient utilization of the student's time. It should be possible through simulated experiments to alter substantially the character of the laboratory work by first letting the student develop an understanding of factors affecting the outcome of the experiment and then, with little need for extensive directions, asking him to design and carry out his own work.

The optimal mixture of individualized instruction by computer and traditional classroom teaching methods remains to be determined, but it is probable that a substantial fraction of normal class time could be profitably replaced by computer-based teaching. In this chapter some efforts to devise computer-based teaching programs as an effective learning aid to be coupled with both lecture and laboratory material will be outlined.

## II. PLATO COMPUTER-BASED TEACHING SYSTEMS

The terminal, which is the communication link between the student and the computer, is a key component of a computer-based teaching system. Many of the terminals in current use involve a serial output, as with a teletype; other terminals display material on a cathode [7] ray tube or a screen on which two-dimensional displays may be created through random access to any coordinate of the display. The capability of random access to any portion of a screen significantly increases the types of material that can be generated.

There are a number of ways to provide a suitable screen with full graphic capability on which the student and his tutor, the computer, can work. The PLATO III system [8] (Fig. 1), used to create many of the figures shown here, produces a display on a storage tube and then transfers the image to a television set for viewing. Provision to mix signals from a slide scanner makes it possible to superimpose photographic images on the computer-generated information in the TV output. Although this system provides considerable flexibility, it suffers from the nonlinearity and low resolution of typical TV monitors. In addition, current versions are relatively expensive to maintain.

The PLATO IV system [2] currently being developed at the University of Illinois by D. L. Bitzer and co-workers replaces the cathode ray tube with a newly developed plasma panel. This panel consists of a matrix of 512 by 512 digitably addressable points within an 8-1/2 x 8-1/2 in. flat glass container [9]. Upon application of appropriate voltage a discharge at any point can be induced in the neon-containing gas mixture. The inherent memory, brightness, and high resolution of this display provide a very convenient

Fig. 1. A PLATO III computer-based teaching terminal.

communication link between students and computer. Since the plasma panel is both flat and transparent, rear-screen projection of slides which become superimposed on the computer graphics generated in the panel is provided.

This type of display system, a PLATO IV student terminal [10], is illustrated in Fig. 2. With both PLATO III and IV, the student interacts with the display through a keyset which resembles an electric typewriter; his work appears on the screen as he types.

The effective utilization of a teletype, cathode ray tube, or plasma display is greatly facilitated by coupling it to a computer system designed to meet the needs of interactive instruction. These needs include essentially instantaneous response to the student's answer to a question, request for help, etc., and the ability to rapidly generate a display on the screen. With the PLATO system, for example, response times of 0.1 sec are typical. The writing rate on a plasma panel of 180 characters/sec and 60 connected lines/sec permits complex displays to be generated in an acceptable period of time. These fast response times and writing rates make it possible for the student to concentrate on the subject matter rather than on the computer system.

Fig. 2. A PLATO IV student terminal with plasma panel.

## III. LANGUAGES FOR WRITING INTERACTIVE INSTRUCTIONAL PROGRAMS

Since an effective instructional computer system should also be designed to meet the needs of the instructor who is to prepare the lesson material, specialized languages [11] have been developed for the preparation of interactive lesson materials. One of these languages, TUTOR [12], which is used with the PLATO system, consists of approximately 200 computer commands, although much lesson material can be written with less than a dozen of these commands. For example, the complete TUTOR program required to present the problem of suggesting a reagent to convert a

2-butanol to 2-butyl chloride and judging the correctness of the answer is given in Table 1.

TABLE 1

PLATO III Computer Program

| | |
|---|---|
| UNIT | ROH |
| WHERE | 401 |
| WRITE | Indicate a suitable reagent for this conversion. |
| | $CH_3CH_2CH(OH)CH_3 \longrightarrow CH_3CH_2CH(Cl)CH_3$ |
| ARROW | 1315 |
| SEN | ($SOCl_2$, $PCl_3$, $PCl_5$, HCl*$ZnCl_2$, thionyl*chloride, phosphorus*trichloride, phosphorus*pentachloride |

The first line, UNIT ROH, simply gives this problem the arbitrary name ROH so that the computer can locate it. The command WHERE followed by 401 instructs the computer that the following sentence is to be written on line 4, beginning at space 1, of the student's screen. The ARROW 1315 places an index arrow on the screen on line 13, space 15, and designates where the student's work, communicated through the typewriter keyset, will appear on the screen. The instruction SEN is a general statement which allows the system to interpret a sentence typed by the student. In this case, a list of suitable reagents is indicated. Note that words that should be coupled, such as zinc chloride, are simply indicated by a * between them.

In the PLATO system, the programs are written on-line and are available for immediate testing. A photograph of the computer program as it appears to the author on the TV monitor is shown in Fig. 3. The problem as presented to the student is shown in Fig. 4.

## IV. RESPONSE OF PROGRAMS TO THE STUDENT

A minimal capability of a computer-based educational system is that it is able to indicate, essentially instantaneously, whether the suggested answer is correct. This capability is illustrated in Fig. 5, where water, the suggested answer to the question, has been crossed out and judged NO. Although the ability to render this simple judgment within 0.1 sec can represent a substantial improvement over commonly employed teaching

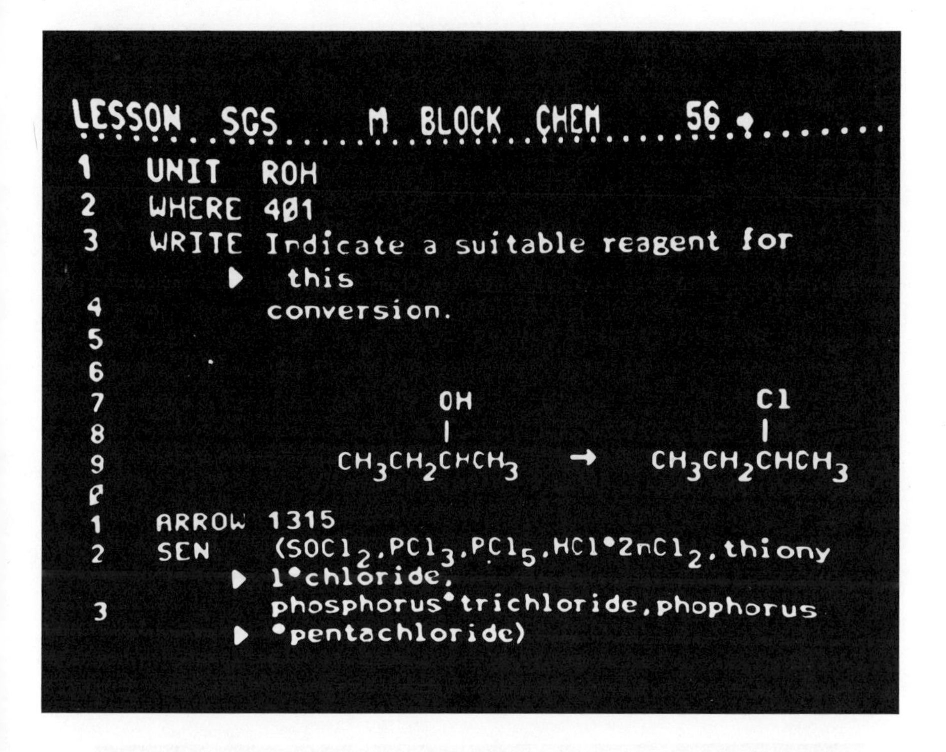


Fig. 3. Program described in Table 1 as it appears to the author on the PLATO III system.

methods, the speed and decision-making capacity of a digital computer makes even more detailed and helpful responses possible. For example, a simple expansion of the program also provides a suggested correct answer, as illustrated in Fig. 6.

If the student tries a wrong answer, the computer immediately gives the correct answer, which must then be typed in before new material is presented. In use this format tends to be too easy for the student. A simple alternative is to give hints on the first error, and only after repeated tries which clearly demonstrate that the problem has been given some thought and cannot be answered will a correct answer be given. In use this scheme can lead to considerable frustration for the student, since he cannot move to a new problem until this one has been solved and there is no help available until at least something has been tried. Thus, it seemed desirable to further modify this type of programming by providing the student with a way out of his difficulty. In this case, the addition of the statement "For help press HELP" at the bottom of the

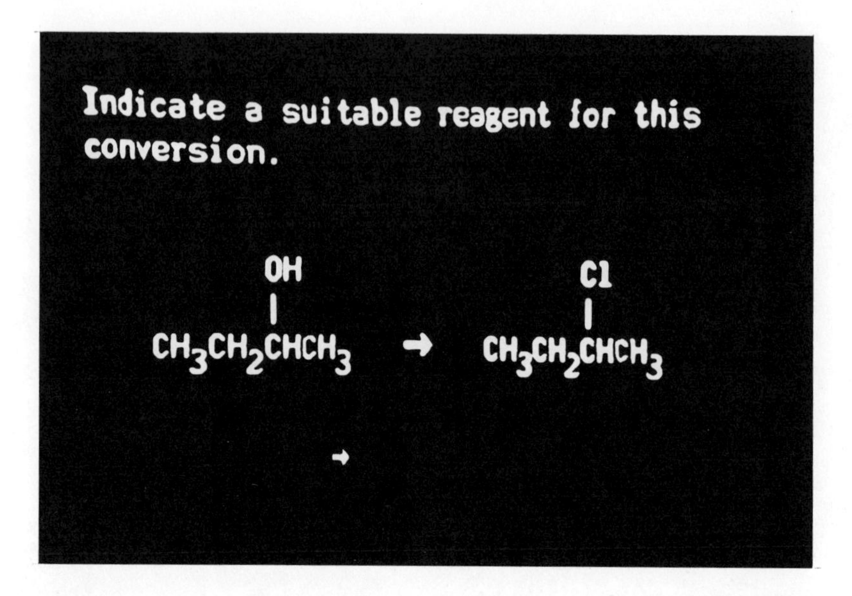


Fig. 4. Program given in Fig. 3 as presented to the student on the PLATO III system.

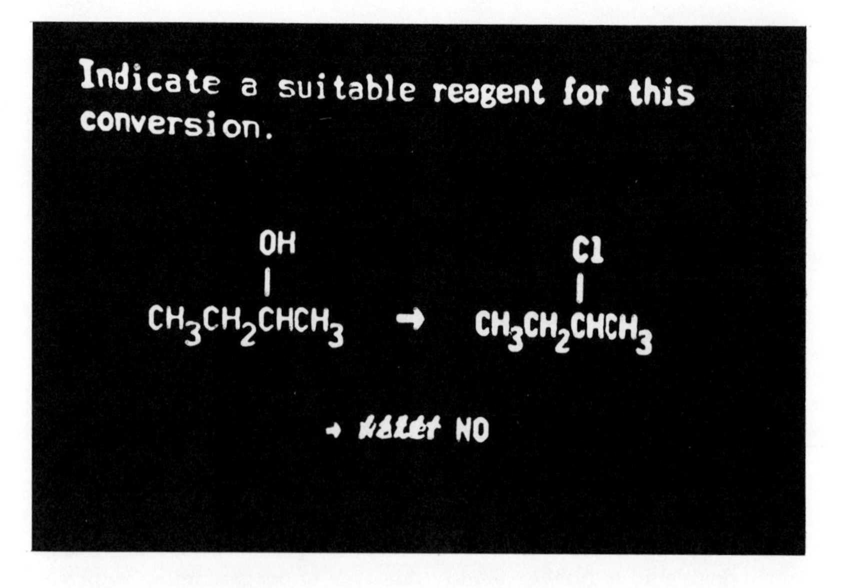


Fig. 5. The suggested answer, water, has been judged NO.

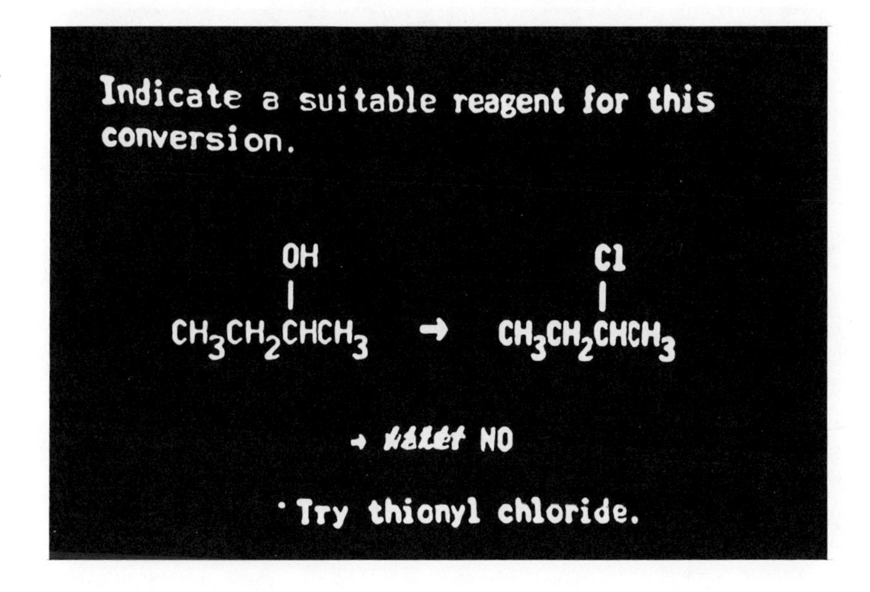


Fig. 6. A modified version of the program given in Fig. 4 suggests a suitable reagent in response to an incorrect answer.

screen allows the student to request assistance. The help supplied in the student-selected branch in the program could be, for example, a general discussion of the chemistry of this type of reaction.

It is, of course, also important that an answer in fact be wrong before remedial comments or help are provided. Unnecessary words such as "I would use...." must be accepted along with blank spaces and capital letters. "Thionyl Chloride" is fully as correct as "thionyl chloride," and the computer must be programmed to accept both answers. Common typing or spelling errors require special consideration, and algorithms are available which automatically detect such errors. Thus, as illustrated in Fig. 7, "thinyl" is underlined and the computer has suggested that the spelling be checked. The expanded program, written in the TUTOR language used with PLATO III, which incorporates branching by pressing HELP, spelling checks, and graduated help, is given in Table 2.

The command HELP in this program instructs the computer system to present a help sequence on the preparation of alkyl halides if the student presses the key marked HELP. When the help sequence has been completed, the student is automatically returned to the point in the program at which he requested help. The simple statement SPELL is all that is required to check automatically for spelling errors, and DIDDL indicates that additional words in the answer are permissible. If a correct response

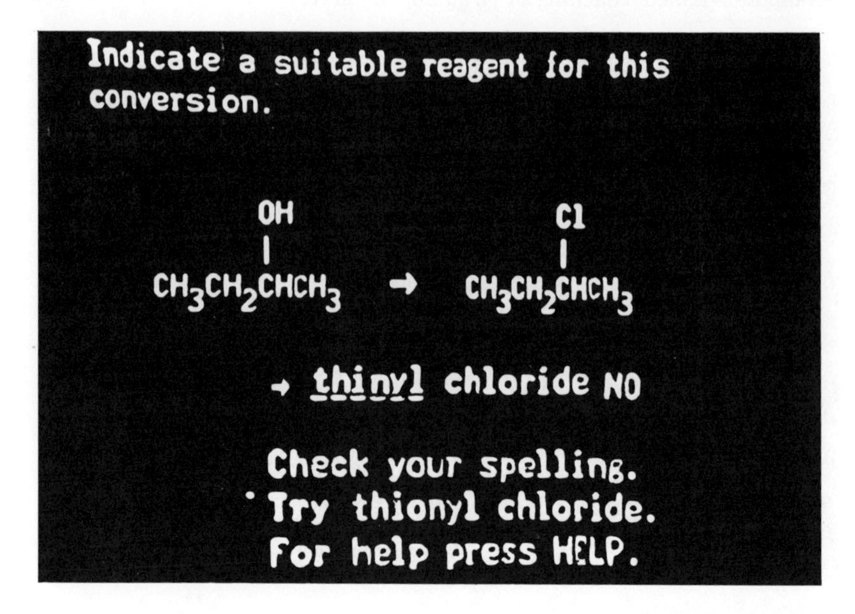


Fig. 7. The incorrect spelling of thionyl is automatically indicated by underlining the "thinyl." The comment "check your spelling" emphasizes the nature of the error.

as determined by the SEN statement is not given, the answer is judged NO. The WRITC statement following CANT instructs the computer to write a comment contingent on the value of I 1 when a response is judged NO. The counter I 1 is set at zero at the start of the problem and increased by one for each error. Thus, in this example, no comment other than NO is given for the first wrong response. The second error results in the message "Try a reagent...." and, finally, a correct answer, "Try thionyl chloride," is given after the second error.

Flexibility in judging the correctness of numerical responses is also necessary for the effective utilization of computers in teaching many aspects of chemistry. In the example illustrated in Fig. 8, any answer within $\pm$ 10% of 2.1 liters would be acceptable, as would the equivalent volume expressed in milliliters, etc.

With the PLATO III system these requirements are easily met with the following portion of a program. The instruction PUTS looks for the units

TABLE 2

PLATO III Computer Program

| | |
|---|---|
| UNIT | R'OH |
| HELP | Halides |
| ZERO | I1 |
| WHERE | 501 |
| WRITE | Indicate a suitable reagent for this conversion. |
| | $CH_3CH_2CH(OH)CH_3 \rightarrow CH_3CH_2CH(Cl)CH_3$ |
| WHERE | 1715 |
| WRITE | For help press HELP. |
| ARROW | 1315 |
| SPELL | |
| WHERE | 1515 |
| WRITE | Check your spelling |
| DIDDL | |
| SEN | ($SOCl_2$, $PCl_3$, $PCl_5$, HCl*$ZnCl_2$, thionyl*chloride, phorphorus*trichloride, phosphorus*pentachloride) |
| CANT | |
| WHERE | 1615 |
| WRITC | I 1,,,Try a reagent which converts the OH into a good leaving group and is a source of chloride ion., Try thionyl chloride., |
| ADD1 | I 1 |

of volume in the student's answer and changes his unit to the appropriate conversion factor:

```
PUTS      milliliters = x 1
          ml = x 1
          cc = x 1
          liters = x 1000
          l = x 1000
STORA     Fl
ANSV      2100, 10% Fl
```

For example, 2 liters becomes 2 x 1000. The instruction STORA computes the value of the resulting arithmetical expression and stores the result in a storage register designated F1. The answer, ANSV 2100, 10%, F1, indicates that the correct response is 2100 $\pm$ 10% and that the student's answer can be located in storage register $\overline{F1}$.

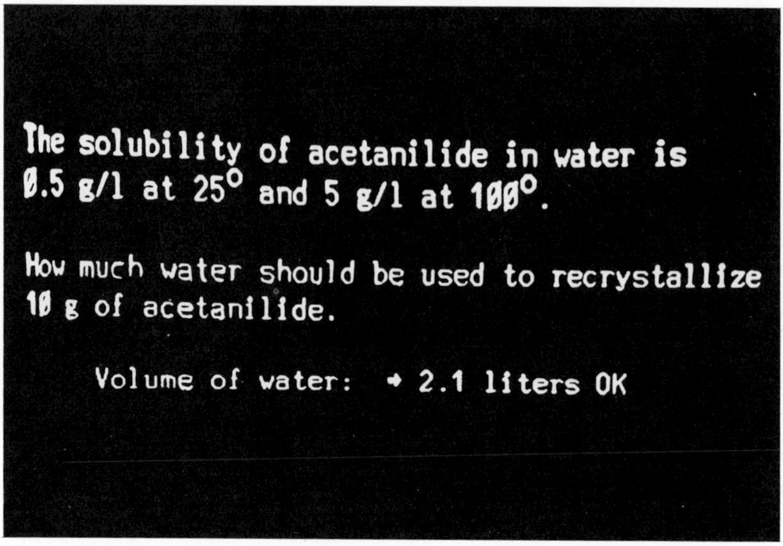

The solubility of acetanilide in water is
0.5 g/l at 25° and 5 g/l at 100°.

How much water should be used to recrystallize
10 g of acetanilide.

Volume of water: ➔ 2.1 liters OK

Fig. 8. Example of computer program in which numerical answers are judged within stated limits with automatic evaluation of the units.

With many types of lesson materials, experience indicates that a limited set of wrong or generally unacceptable answers will be encountered with high probability. Designing the program to look for these specific errors and then to provide detailed or specific comments can be quite effective. This is illustrated in Fig. 9, where the explanation of the student's error, "You need a hyphen after the i," has been given to the student. In the TUTOR language used for PLATO III, this is accomplished as follows:

```
CANT        i pentane
WRITE       You need a hyphen after the i
```

Thus, in this instance the student is required to follow standard rules of nomenclature, but he is not misled or irritated by having a response he considered valid rejected.

The requirement that all reasonable correct answers be accepted is easily dealt with by simply trying the material on a group of students. The PLATO computer system provides the instructor with printed records of each student's work and how each entry was interpreted. Review of these records after each use coupled with appropriate modification of the program logic readily produces lesson material which is quite free of the complaints which inevitably result when a correct response is judged "NO."

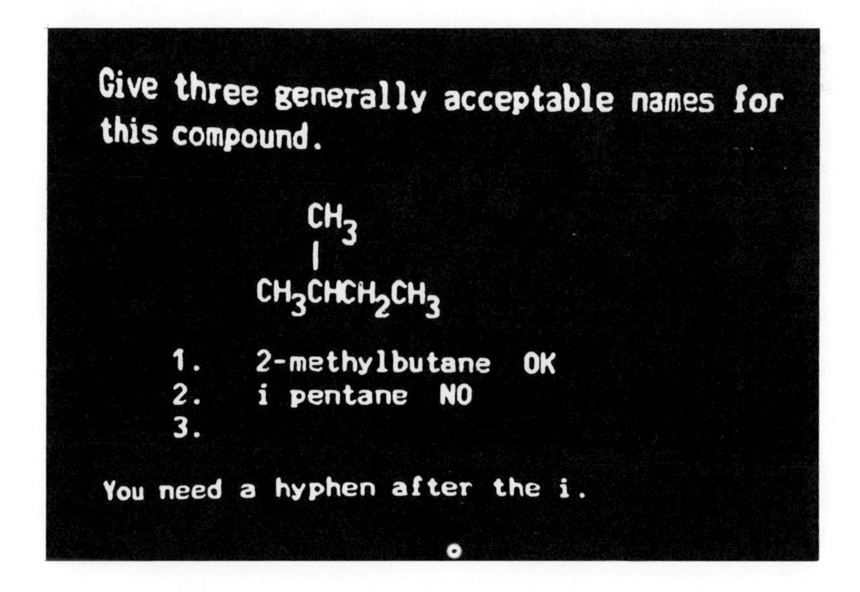


Fig. 9. Common errors may be indicated in the program so that specific comments are given on the screen when they occur.

This type of computer programming, done directly on-line at a student terminal and viewed and tested immediately, is simple enough that teachers with no prior knowledge of computer programming can rapidly learn to produce innovative instructional material. As will be illustrated later, complex numerical problems and dialogue involving large vocabularies are also readily handled.

Given the fact that a computer can write, draw, show slides on the screen, and respond to an input from the student through the keyset, the central problems become what to put on the screen, what to do if the student's response to a question would be considered correct, and what to do if the response is either not interpreted by the computer or is simply wrong. Unfortunately, it is still difficult to provide documented procedures to optimize both learning and student acceptance.

## V. BASIC LESSON STRATEGIES

Several overall strategies are possible in developing material on a specific topic or concept [13]. For example, a tutorial approach, presenting concepts by posing a logical series of questions or problems, with built-in help, can be effective. However, students like to have the option of deciding themselves when review is desirable. One way to provide this option is to divide the program into segments in which records on performance are maintained by simply incrementing a counter every time an error is made. At logical points in the development of the material the student is given the opportunity to review. Of course, performance records may dictate an automatic review of the material until performance has reached acceptance levels.

Reviews can take many forms. For example, a given problem can be presented repeatedly, regularly inserted into a series of other material until it has been mastered, or additional related problems can be given

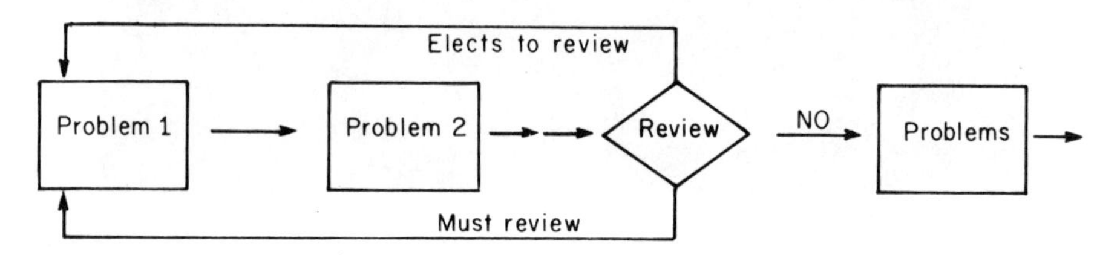


until the error rate becomes acceptably low. However, it is important not to force a student into a lengthy review sequence on the basis of a single error. Combining student-elected review with automatic review based on performance on several related problems provides both a measure of freedom of choice and the assurance that the material has been mastered. An example of a page summarizing some possible review sequences taken from a lesson on distillation is illustrated in Fig. 10. The list of topics to review increases in length automatically as the student progresses through the lesson.

Another interesting overall structure for a lesson gives the student the option of working through background material or going directly to a series of problems. Here it is extremely important to indicate to the student at

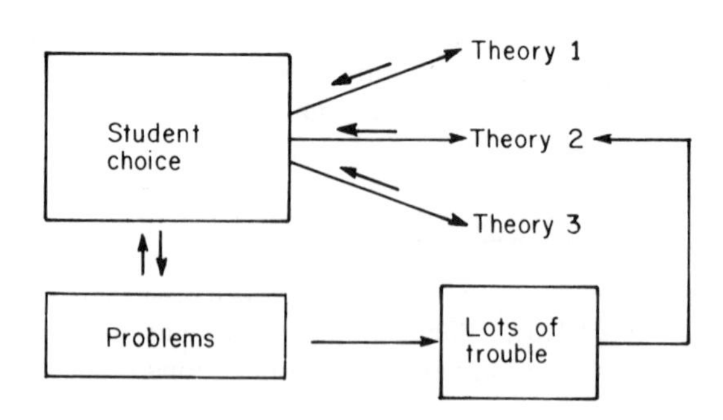

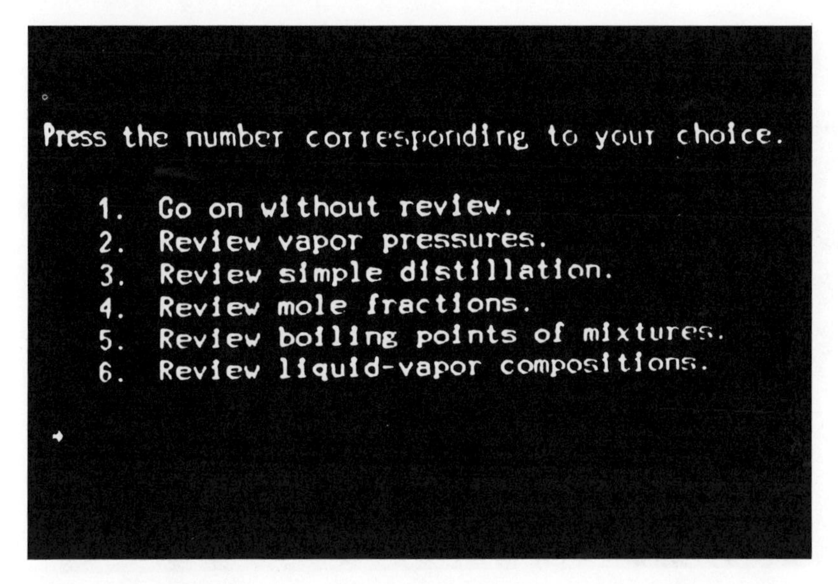


Fig. 10. Example of a review choice page part of the way through a lesson on distillation.

each stage that the choice is reversible. This structure makes it easy to test skills and then, if necessary, review the development of the theory and the typical student reviews many times. Since error rates and review selection can be recorded, a student who is having difficulty but does not review can be further advised of the option and, if necessary, given additional help. Because of the flexibility of computer-based educational systems, many types of review structures may be included in a lesson.

## VI. APPLICATIONS OF LESSON STRATEGIES

### A. Ketone Chemistry

Each problem or lesson sequence in these overall lesson strategies can be quite different from those used in conventional teaching situations because of the use of a high-speed digital computer. For example, a given set of concepts can be established by providing a mathematical model of some chemical system which the student can manipulate and explore, as illustrated in Fig. 11, which is a portion of a program on ketone chemistry. The student is to study the rate of bromination of a ketone as a function of both bromine and base concentration. Here a plot of yield vs time is provided [14] for the student in response to his choice of concentration of base and

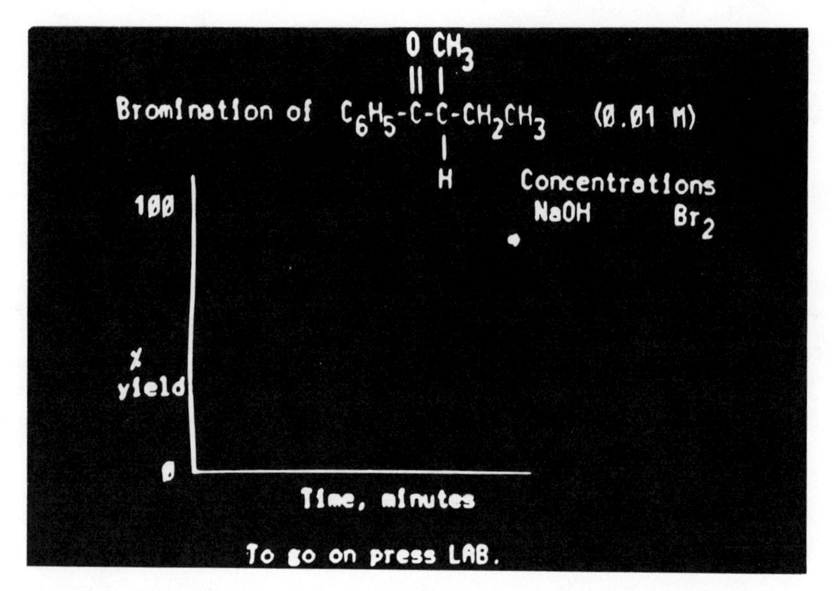


Fig. 11. Photograph of display before an investigation of the effect of the concentration of bromine and base on the rate of bromination of ketones is begun.

bromine (Fig. 12). Any number of experiments can be simulated. When he is satisfied that he understands the relationship between concentration of base and bromine and relative rate, the student can elect to go on, at which point he is asked questions (Fig. 13) testing his understanding of the factors affecting the reaction. It is clear from Fig. 13 that these questions require the student to formulate an answer in his own words. Techniques of phrase-judging now available make it quite easy to interpret a given answer within the context of the subject matter, and the clues inherent in multiple-choice responses are thus avoided.

In this particular problem, if the correct answer is not given, experiments are suggested which will, if properly interpreted, yield the desired information. Of course, each experiment requires only about 0.5 sec. It should be noted that this approach expands the initial objectives of establishing experimental facts about the reaction to include the design and interpretation of experiments.

## B. Spectroscopy

Presentation of nculear magnetic resonance spectra is another example of the computer-generated simulations of specific chemical systems used

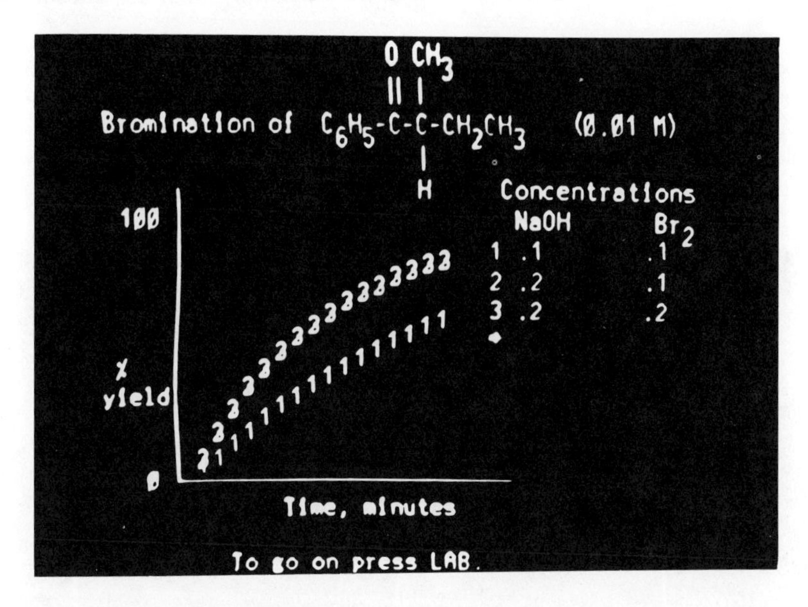


Fig. 12. Results of three experiments plotted as yield of product vs time. Experiments 2 and 3 give the same rate.

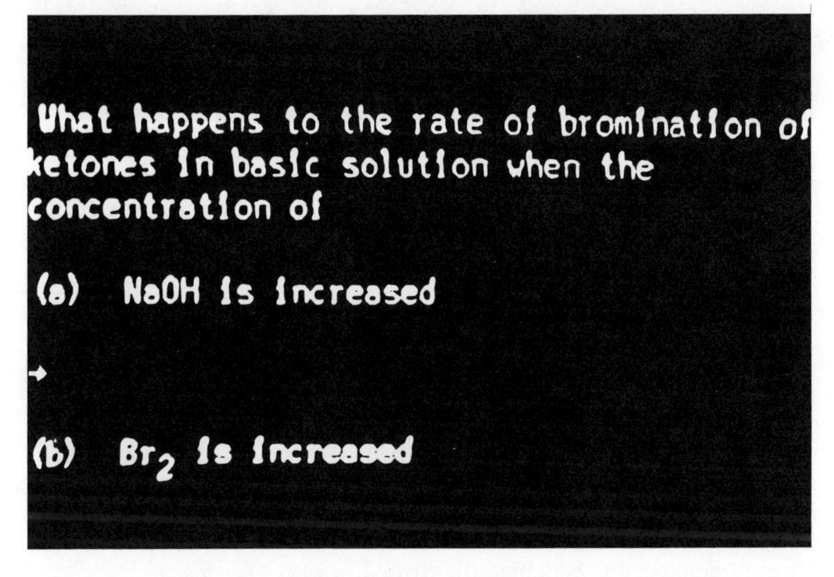


Fig. 13. If the correct experiments in Fig. 11 and 12 have been done, these questions can be answered. Appropriate experiments are suggested in response to incorrect answers.

as part of a lesson sequence. Since NMR spectra of simple systems can be calculated on-line [15] and displayed on a student's screen (Fig. 14) he can examine the relationships between the appearance of a spectrum and the values of coupling constants and chemical shifts. When he is satisfied with his progress, a tutorial interaction can be employed to assure that an adequate understanding has been achieved.

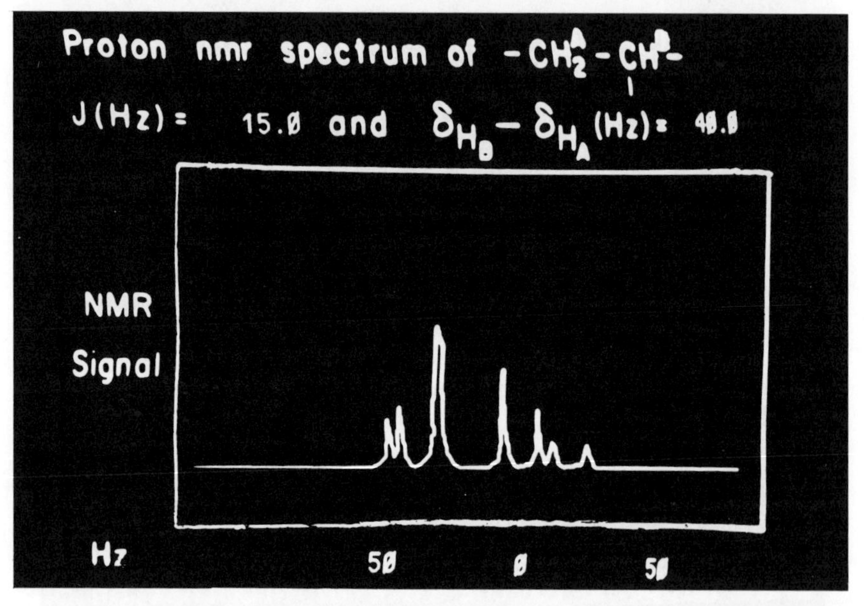


Fig. 14. Part of a lesson on NMR spectroscopy. Here the student studies the effect of J and δ on the appearance of the spectrum by indicating values for these parameters. The computer responds by calculating and displaying the resulting NMR spectrum.

## C. Distillation

In addition to presenting a static display for the student to respond to, it is possible to confront him with a dynamic situation which requires that decisions be made in the appropriate time relationship. This concept is illustrated by one segment of a lesson on distillation which requires the student to interact with the lesson as the display depicting the course of the experiment develops. This particular example concerns a portion of an exchange between student and computer in the fractional distillation of a mixture of pentane and heptane, illustrated in Fig. 15.

As the experiment proceeds, the computer provides a graph of the temperature at the top of the distillation column vs milliliters distilled.

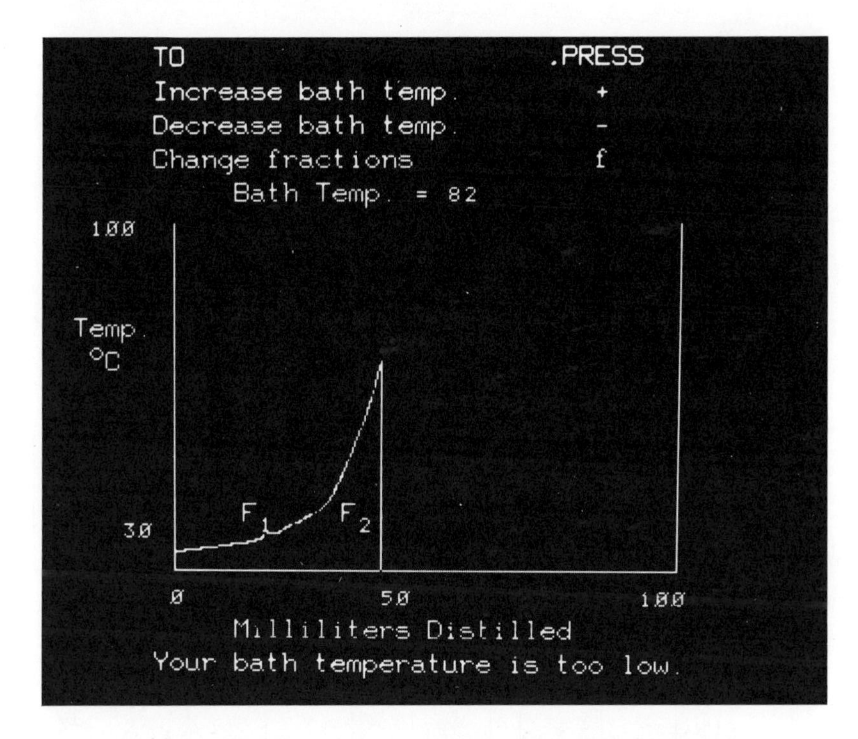


Fig. 15. Portion of a simulated fractional distillation.

The student must control the rate of distillation by periodic adjustment of the oil-bath temperature. If the bath temperature is too low, distillation stops. If the temperature is too high, the student is advised of his error and is asked to start the experiment over again. The points where the student decided to change fractions are marked as Fl, F2, etc. The summary and analysis of the experiment (Fig. 17) indicates that fraction 1 is not acceptable because not enough material was collected in the first fraction. The experiment must be repeated. Because of time, space, and equipment limitations, repeating actual laboratory experiments is generally not possible. One additional advantage of computer use is that, in just a few minutes, the interested student can design his own distillation problem and test his skill at solving it.

A general outline of the procedure used to develop the distillation simulation is given in Fig. 16. First an appropriate initial bath temperature is selected. The lesson then checks the value suggested and provides diagnostic feedback to the student. Next, the display showing the coordinates in the graph is put on the screen. If the student presses PLUS the program

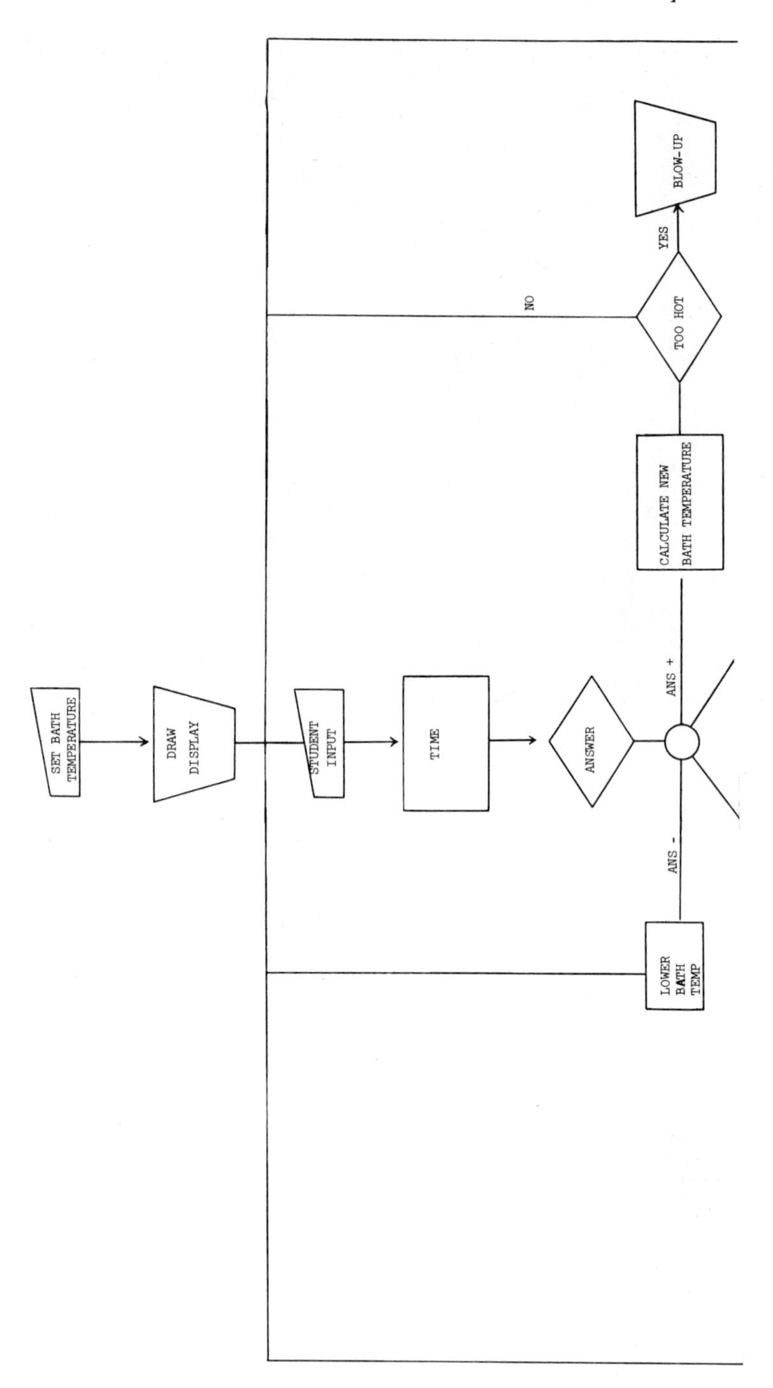
SET BATH TEMPERATURE
DRAW DISPLAY
STUDENT INPUT
TIME
ANSWER
ANS +
ANS -
CALCULATE NEW BATH TEMPERATURE
TOO HOT
YES
NO
BLOW-UP
LOWER BATH TEMP

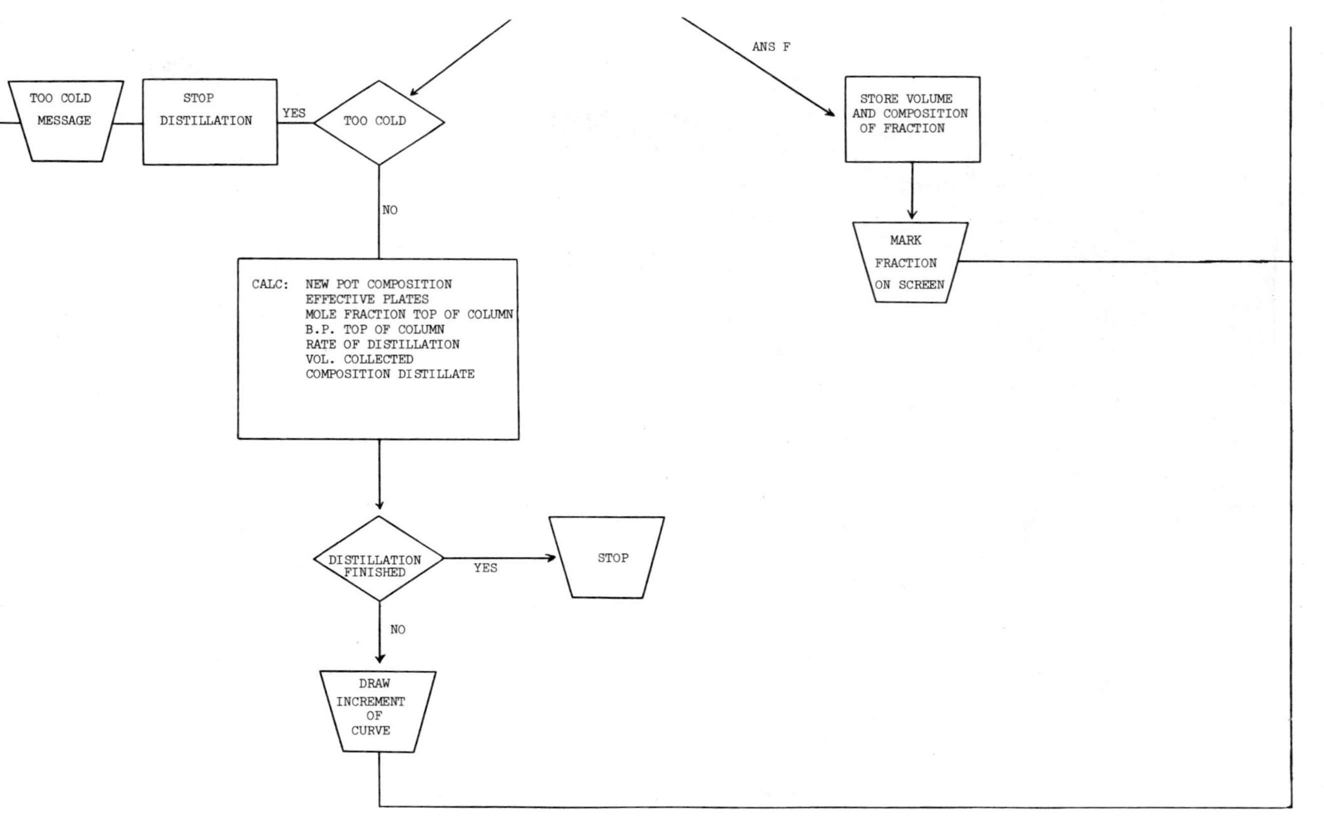


Fig. 16. Flow chart illustrating computer logic used in fractional distillation sequence.

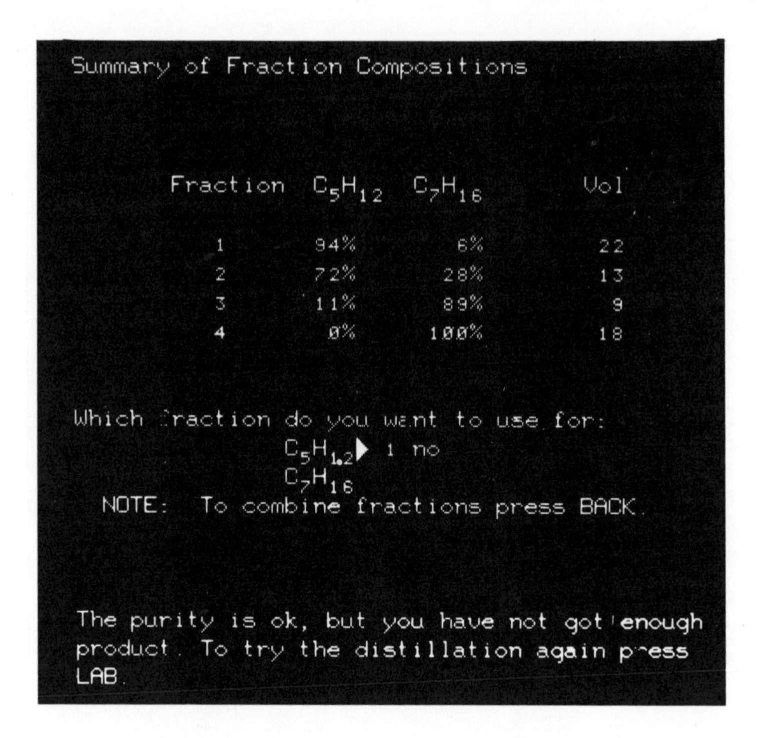


Fig. 17. Summary of a typical distillation experiment.

branches to the necessary routines to set the new higher bath temperature and increase the distillation rate, and then checks to assure that the temperature isn't so high as to cause excessive flooding in the distillation column. Pressing MINUS correspondingly lowers the bath temperature and pressing F collects a fraction, marks the point on the graph of bp vs ml distilled, and stores information on volume and composition. If no input from the student is received within the allotted time, which is coupled to the distillation rate, the progress of the distillation is incremented one drop, provided the bath temperature is high enough to maintain reflux. Incrementing the distillation rate involves calculating the new composition of the material in the pot, the number of effective plates, which is related to the rate of distillation, the mole fraction of the components at the top of the column and, assuming ideality, the boiling point at the top of the column [16]. The new distillation volume and composition must, of course, also be calculated. If all of the material has not been distilled, an increment of the bp vs volume plot is drawn on the student's screen and the process is repeated.

## D. Identification of Unknowns

Another type of lesson structure involves a simple dialogue between the student and the computer on a specific topic. Such exchange, with a minimum of artificial structure, tends to maximize flexibility for the student. A conceptually simple example of this type of program is one in which the student must identify [17] an unknown compound by asking questions about it. For this technique to be successful, it is important that the computer recognize and respond to a large vocabulary so that the student can express a question in his own words [18]. In a lesson of this type a vocabulary of 1800 words and phrases is sufficient to respond appropriately to better than 90% of the questions presented by a class of 14 students over a two-hour period. On PLATO III the machine time required to understand the question and locate the answer is in the range of 25 to 50 msec. In use, the student simply asks a series of questions and receives an essentially instantaneous response. Typical questions might be, "Where does it boil," "Show me the infrared spectrum," "Does it make a derivative," etc. On the average it takes 20 min for a student using this approach to identify an unknown on the computer.

The TUTOR language for the PLATO IV system makes writing lessons of this type particularly simple [18]. The general structure of the program is given in Table 3. In this program, a vocabulary for the particular subject matter is typed into the lesson (Table 3). Words enclosed by < > are ignored by the computer and words enclosed by ( ) are considered synonyms. The ideas or concepts to which the lesson is to respond are then listed using any of the vocabulary words. The answers to questions dealing with each of these ideas are written into the program in the same order as the concepts. The statement WRITEC selects and writes the answer on the screen, contingent on the position of the question in the list. Although many other structures for organizing such a lesson can be envisioned, the essential point here is that the computer-based educational system, in this case PLATO IV, provides the classroom instructor with a format for writing the program which makes it possible to focus attention on the subject matter rather than the mechanics of computer programming.

## E. Recrystallization

A further conceptual development of a dialogue between the student and computer on some topic makes the answer to a question a function of previous exchanges. One example of this is illustrated in Table 4, which summarizes part of a lesson on crystallization. In this case the student

TABLE 3

Qualitative Organic Chemistry Dialogue

| | |
|---|---|
| VOCAB | < a, an, and, can, could .... > |
| | (odor, smell, smelly, fragrance, aroma) |
| | (soluble, solubility, dissolve) |
| | (mp, melt, melts, melting, freeze, freezing, solidify) |
| | . |
| | . |
| | . |
| WHERE | 501 |
| WRITE | Type your question about the unknown and then press NEXT |
| ARROW | 810 |
| CONCEPT | odor |
| CONCEPT | dissolve in water |
| CONCEPT | Where does it melt |
| . | |
| . | |
| . | |
| WRITEC | Ansent, |
| | It has a strong aromatic odor, |
| | It is slightly soluble in water., |
| | The melting point is below $0^{\circ}$ C., |
| | . |
| | . |
| | . |

must describe how he would proceed in the laboratory to recrystallize naphthalene from methanol. The response to the student's request that some methanol be added to the naphthalene is, of course, a function of what he has done previously. In the example illustrated in Table 4, PLATO has suggested that the sample be weighed first. As this exchange proceeds, the computer must calculate the response to the student's question based on sample size, temperature, etc. For example, if the solution is cooled, the product might crystallize, form an oil, or stay homogeneous, depending on what had happened previously. Combining a dialogue between the student and the computer with a mathematical algorithm of a chemical system to determine the character of the answer results in a responsive and instructive lesson sequence.

TABLE 4

Portion of a Dialogue between PLATO and a Student Attempting to Recrystallize Naphthalene from Methanol

| | |
|---|---|
| Student: | Add some $CH_3OH$ |
| PLATO: | Before adding methanol you should weigh your sample |
| Student: | OK, how much does it weigh? |
| PLATO: | The sample weighs 22 g |
| Student: | Where does it melt? |
| PLATO: | The mp is 72°-78°, but it doesn't all melt |
| Student: | Now add methyl alcohol |
| PLATO: | It is a good idea to save seed crystals before adding solvent. Some compounds are difficult to crystallize without them. |
| Student: | Save seed crystals |
| PLATO: | OK, a few crystals have been put in a small bottle |
| Student: | Let's try again, add some $CH_3OH$ |
| PLATO: | How much methanol would you like to add? |
| Student: | 40 ml |
| PLATO: | About 10% of the sample dissolved in 40 ml of methanol at 20° C |
| Student: | Put it on a steam bath |
| PLATO: | Almost all of the material dissolved |
| . | |
| . | |
| . | |

## F. Synthesis

The concept of a lesson that computes the correct response based on a mathematical model and the parameters supplied by the student is further illustrated by multistep organic synthesis programs. Here the student must provide a synthesis of an organic compound from suitable starting materials. Since such problems often have a large number of possible solutions, attempts to anticipate all of them with pre-stored answers rapidly becomes difficult to handle. The lesson illustrated in Fig. 18 is just an extension of the concepts illustrated in the qualitative analysis and recrystallization programs. In the synthesis lesson, the vocabulary consists of the organic chemist's collection of reagents and starting

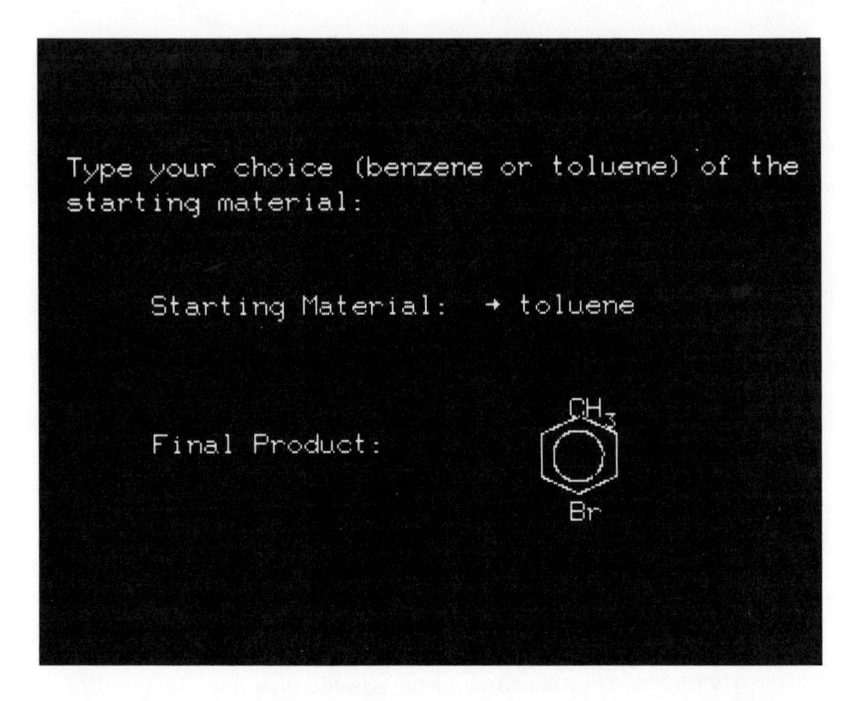


Fig. 18. Selection of starting point in synthesis of benzene derivatives.

materials. The response to the student's suggestion of a particular reagent is computed based on the appropriate functional group chemistry [19].

This approach is illustrated for electrophilic aromatic substitution in Fig. 18-20. In this example, the student proceeds by selecting his starting material. The computer responds by asking for the reagents to be used in the first step of the synthesis. Although the choice is not optimal, nitration has been suggested as the first step. PLATO has responded by showing the composition of the mixture resulting from this reaction (Fig. 19). Selection of the appropriate isomer, followed by the additional steps indicated, yields the desired product. The entire synthesis, as proposed by the student, is summarized in Fig. 20. Of course, many other syntheses are possible. At each point in the synthesis, checks are made to ensure that the proposed reaction conditions are compatible with the molecule to be constructed by the student.

At each step in the synthesis a check on the yield is made. If the amount of product is too low, the proposed synthetic route is terminated with a simple message suggesting the use of another route. If the yield is adequate, then the product of the preceding reaction is compared with the desired

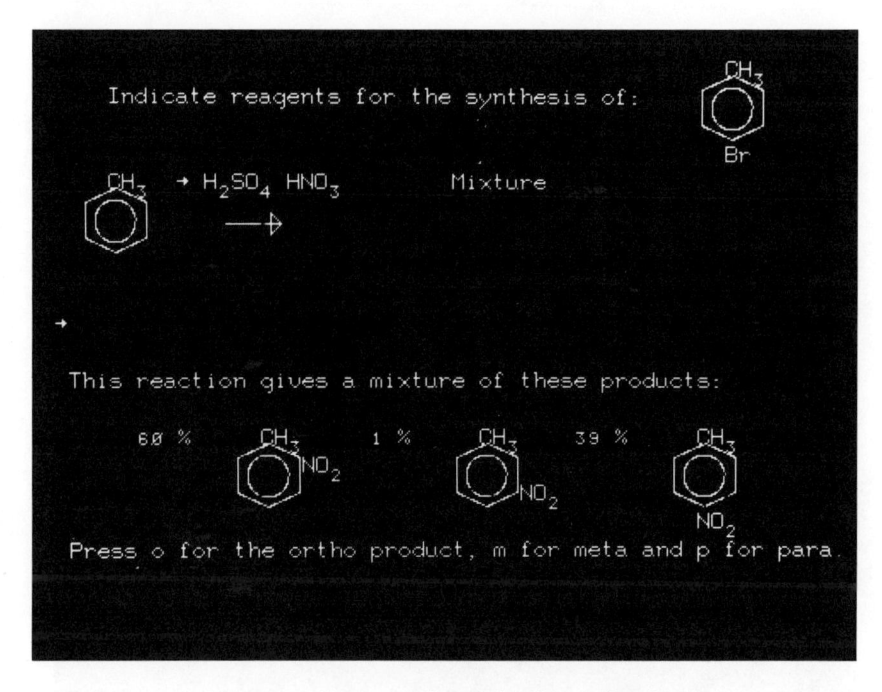


Fig. 19. First step in student's synthesis.

target molecule. If they are the same, the synthetic scheme is terminated and the choice of either trying another scheme for the same product or going on to a new problem is given. However, if the synthesis is not complete, the computer is instructed that the student will suggest the reagents for another step. After a new reagent has been entered, the program checks to see if it is known to PLATO. If the reagents are recognized, the program is directed to that branch of the instructions dealing with the particular reaction. Basically, there are two types of reactions possible in aromatic synthesis, either introduction of a new group or modification of an existing group.

In the case of nitration, for example, it is first necessary to look up appropriate rho values and steric parameters [20]; actual literature values are used whenever possible. In order to calculate substitution orientations, the number of substituents already on the ring is determined and the appropriate branch in the program is taken. Checks for incompatible groups and adequate electrophilicity for the reactants are next made. A linear free-energy calculation, including steric effects [21], is made to determine isomer distribution, followed by the appearance of the structure of the products

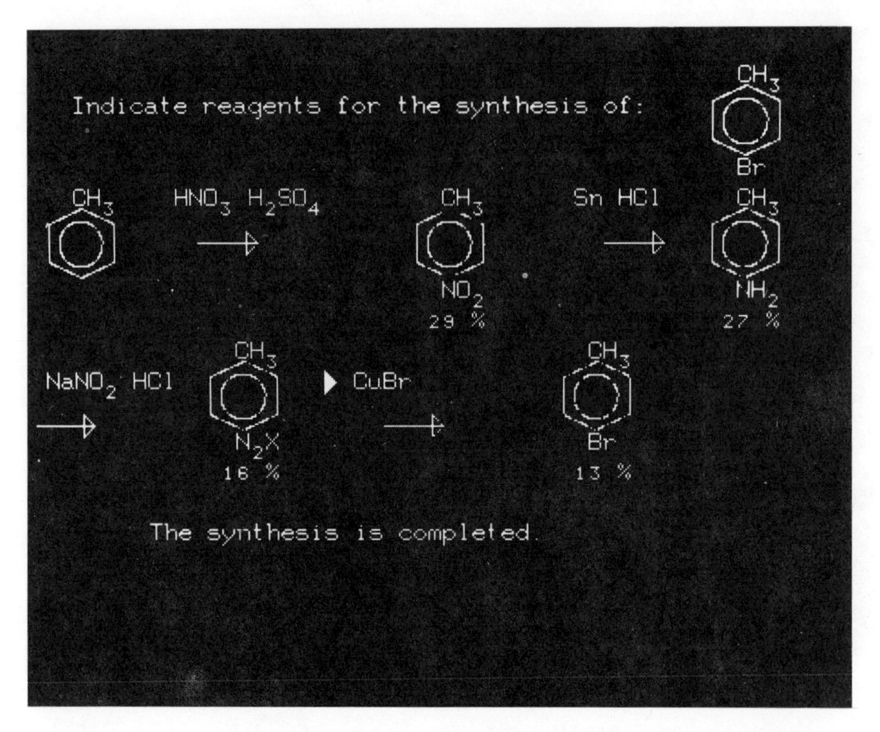


Fig. 20. Completed synthesis. Alternative routes which may give higher yields may be explored or another problem can be requested.

on the student's screen. Upon selection of one isomer by the student, the process is repeated. Aliphatic synthesis has been done in a similar format.

## VII. EVALUATIONS

Only a few of the techniques for presenting chemistry made possible by interactive computer-based teaching methods have been outlined here. Furthermore, expanding technology and experience are rapidly increasing the capabilities of computer-based teaching systems. With the technical capability to develop complex interactive lesson materials, it is interesting to compare teaching effectiveness of such lesson materials with conventional lectures, discussions, and laboratory work. Although such comparison often rests on the assumption that the same instructional objectives obtain for both types of presentation, instructional goals are nonetheless closely coupled to the particular teaching methods being used, and having a large digital computer as an assistant often induces the instructor to change the content of the course to take advantage of the available technology. This perturbation of teaching objectives with the method of instruction requires that evaluation be done with unusual care and that the content of computer programs not be artificially restricted in order to make them conform to the content of conventional course materials or a particular evaluation procedure [22].

Since the application of computer-based teaching is just developing and facilities for performing large-scale experiments with statistically significant numbers of students are usually not yet readily available, studies on effectiveness are fairly limited in scope. However, recent investigations by Rodewald, Culp, and Lagowski [4] and by Grandey [5] indicate strong acceptance by students of the concept of computer-based teaching and suggest improved performance in conventional examinations by students who have used related computer-based lesson materials. Thus, it seems apparent that the ability of computer systems to present complex interactive lessons and to gather and correlate data on the performance of individuals as well as the teaching effectiveness of particular techniques and programs has provided the tools by which the quality and effectiveness of instructional techniques can be significantly improved.

## ACKNOWLEDGMENTS

This work was made possible by the cooperation and assistance of Professor D. L. Bitzer, Director of the Computer-Based Educational Research Laboratory at the University of Illinois, his staff, Mr. Paul Tenczar, who developed many aspects of the programming procedures used, and Professor D. Y. Curtin. The Computer-Based Educational Research Laboratory is supported in part by ONR NONR 3985 (08), Advanced Research Projects Agency, the National Science Foundation, and the University of Illinois.

## REFERENCES

1. R. Gerard, Proc. Nat. Acad. Sci., 63, 573 (1969).
2. D. Alpert and D. L. Bitzer, Science, 167, 1582 (1970).
3. S. G. Smith, J. Chem. Ed., 47, 608 (1970).
4. (a) S. J. Castleberry and J. J. Lagowski, J. Chem. Ed., 47, 91 (1970); (b) L. B. Rodewald, G. H. Culp, and J. J. Lagowski, J. Chem. Ed., 47, 134 (1970); (c) S. J. Castleberry, E. J. Montague, and J. J. Lagowski, J. Res. Sci. Teaching, 7, 197 (1970); (d) G. H. Culp and J. J. Lagowski, J. Res. Sci. Teaching, 8, 357 (1971).

5. (a) R. C. Grandey, J. Chem. Ed., 48, 791 (1971); (b) S. K. Lower, J. Chem. Ed., 47, 143 (1970).
6. B. Sherwood, C. Bennett, J. Mitchell, and C. Tenczar, Proceedings of the Conference on Computers in the Undergraduate Curricula, Dartmouth College, June, 1971, p. 463.
7. For example, IBM 1500 system.
8. D. L. Bitzer, R. W. Blomme, B. A. Sherwood, and C. Tenczar, Proceedings of a Conference on Computers in Undergraduate Science Education, Illinois Institute of Technology and Commission on College Physics, August 1970, p. 335 (published in 1971, available from American Institute of Physics, New York, N.Y.).
9. R. L. Johnson, D. L. Bitzer, and H. G. Slottow, IEEE Trans. Electron Devices, ED-18, 642 (1971). This plasma panel is commercially available from Ownes-Illinois, Toledo, Ohio.
10. J. Stifle, "A Plasma Display Terminal," CERL Report X-15 (Computer-Based Educational Research Laboratory, University of Illinois), 1970. The terminal is currently manufactured by Magnavox Corporation, Fort Wayne, Indiana.
11. For example, APL, Coursewriter, CLIC, TUTOR, Basic.
12. R. A. Avner and P. Tenczar, "The TUTOR Manual," CERL Report X-4 (Computer-Based Educational Research Laboratory, University of Illinois), 1969.
13. (a) W. F. Hill, Learning. A Survey of Psychological Interpretations, Chandler Publishing Co., New York, 1963; (b) R. M. Gange, The Conditions of Learning, Holt, Rinehart, and Winston, New York, 1970; (c) R. C. Anderson, Rev. Educ. Res., 40, 349 (1970); (d) K. L. Zinn, Rev. Educ. Res., 37, 618 (1967).
14. (a) R. P. Bell and H. C. Longuet-Higgins, J. Chem. Soc., 1949, 636; (b) P. D. Bartlett, J. Amer. Chem. Soc., 56, 967 (1934).
15. (a) J. A. Pople, W. G. Schweider, and H. J. Bernstein, High-Resolution Nuclear Magnetic Resonance, McGraw-Hill, New York, 1959; (b) R. E. Rondeau and H. A. Rush, J. Chem. Ed., 47, 139 (1970).
16. A. Rose and E. Rose, in Techniques of Organic Chemistry, (A. Weissberger, ed.), Vol. IV, Wiley-Interscience, New York, 1951.
17. (a) R. L. Shriner, R. C. Fuson, and D. Y. Curtin, Systematic Identification of Organic Compounds, 4th ed. Wiley, New York, 1956; (b) C. G. Venier and M. G. Reinecke, Abstracts, Conference on Computers in Chemical Education and Research, DeKalb, Ill., June 1971, pp. 9-85; (c) F. M. Hornack, Abstracts, Conference on Computers in the Undergraduate Curricula, Dartmouth College, June 1971, p. 359; (d) G. F. Luterie and J. M. Denham, J. Chem. Ed., 48, 670 (1971); (e) W. Gasser and J. L. Emmons, J. Chem. Ed., 47, 137 (1970).
18. Systems programming making this possible was developed by P. Tenczar.

19. (a) E. J. Corey and W. T. Wipke, Science, 166, 178 (1969); (b) J. B. Hendrickson, J. Amer. Chem. Soc., 93, 6847 (1971); (c) J. B. Hendrickson, J. Amer. Chem. Soc., 93, 6854 (1971); (d) E. J. Corey, R. D. Cramer III, and W. J. Howe, J. Amer. Chem. Soc., 94, 440 (1972).
20. R. W. Taft, Jr., in Steric Effects in Organic Chemistry, (M. S. Newman, ed.), Wiley, New York, 1956.
21. J. Ghesquiere and S. Smith, unpublished work.
22. J. M. Atkin, J. Res. Sci. Teaching, 1, 129 (1963).

Chapter 3

# COMPUTERS APPLIED TO PHYSICAL CHEMISTRY INSTRUCTION

Morris Bader

Department of Chemistry
Moravian College
Bethlehem, Pennsylvania



## I. INTRODUCTION

The computer, one of the most sophisticated pieces of machinery yet contrived by man, was created not so much in the physical image of man but rather in the image of the mind of man. The final goal, the perfect machine, may yet have the ability to be taught or to acquire some other humanistic quality, as, say, the beginnings of reason. While that endeavor is still in progress, the computer has been utilized to attempt the reverse, to train humans in the skill of reason. Education, loosely defined, is the transmission of knowledge. This is accomplished by offering the learner a supply of facts, usually defined in the scientific sense by some operational procedure, and a set of explanations, called theories or laws, which tend to rationalize these observations. The facts in science are not self-evident. They must be coaxed from the world about us by consummate skill, by oft times delicate, and most times exotic, apparatus. The theories, too, in the avoidance of the semantic proofs of the Middle Ages, are now couched in the most elegant forms of mathematical expressions which by their precise meaning allow little possibility of semantic distortion. The chasm between mathematics and language has already reached the point where word questions may have little or no meaning. After an intensive lecture describing the probability equations of Schrödinger concerning the behavior of electrons in atoms there is always some student who asks, "But really, professor, where is the electron?" That question has no precise answer. For many years now chemists have been speaking of a chemical bond as the sharing of a pair of electrons between two atoms. Yet, the simple two-electron molecular orbital wave function description of that bond leads to useless quantitative values, while the Hartree-Fock approach involving the solutions of thousands of integrals leads to values in almost precise agreement with nature [A-1]. Should we abandon the simple physical picture easily grasped by all, and useful for its qualitative information, or should we unequivocally accept the mathematical formalisms which, though correct, leave us without any physical model? This question is still to be resolved.

Through all these frustrations and difficulties the major tool of the educator has remained the printed page. But while it has perfect retention of ideas, the printed page is mute, its errors permanently etched, and its revision must be total. The computer as an educational device appears to have many of the best qualities of the printed page, none of its deficits, and above all some new characteristics in communication and recall which will inevitably necessitate radical changes in our educational system.

## II. APPLICATIONS TO PHYSICAL CHEMISTRY

Physical chemistry is one of those areas ideally suited for interaction with the computer. Physical chemistry may be defined as that branch of chemistry dealing with the physical states of matter. In the broad sense it includes the transformations of matter from one state to another and the energies of those transformations. Traditionally, in the college curriculum physical chemistry has been subdivided into its component areas, i.e., thermodynamics, chemical equilibria, kinetics, statistical mechanics, and quantum mechanics. Topics within these divisions continue to grow at an ever-increasing rate. For example, molecular structure and spectroscopy have branched off from the highly theoretical quantum theory in a more qualitative fashion for the practical applications of the organic and inorganic chemists. Within the past few years we have seen the growth of group theory as a study in itself, as well as ligand field theory. While the number of topics continues to grow, the typical undergraduate college career is maintained as a four-year program. The chemistry faculty have the most difficult task of designing a program which will offer the student a sufficiently comprehensive background so that he will be prepared either to enter industry, to continue in graduate work, or to enter some related field of specialization.

There have been numerous approaches to the problem of fitting an expanding curriculum into a four-year time span. Obviously, some topics must be deleted. The choice may depend on the interests of the students, or the instructor, or the needs of the local scientific community. Much of the higher-level material has been moved down in the curriculum to introductory level courses. Nuclear chemistry and chemical equilibria are now generally handled at this early stage. At the same time, much of the qualitative and descriptive chemistry has been condensed severely or deleted in toto. Within the past few years an attempt was made to dispense with the usual introductory course and substitute in its place the traditional physical chemistry. This approach in its extreme often left the student with very little qualitative chemical knowledge. It later prompted a note to a journal that on a doctoral examination a graduate student expounded on the idea that "AgCl is a greenish-yellowish gas." [A-2]. Like the emperor's new clothes, this remark articulated what so many instructors had observed--that splintering and reorganization of topics will no longer suffice as patchwork solutions to the educational dilemma.

The applications of the computer to the educational environment depend largely on both the needs of the student and the computing facilities available. Universities with very large student populations have a direct need for tutorial and drill programs to supplement instructional time. Some small distinction is made when the computer is used to introduce new material to the student; this is generally referred to as CAI, computer-assisted instruction. The philosophy of the tutorial and the development of techniques were pioneered in the mid 1950s. Strong, an early investigator [A-3], pioneered the development of the teaching machine, a black box with a window through which discussion, questions, and answers were displayed to the student. Except for the manual crank, many computer methods still adhere closely to this procedure. This approach, to be successful in reaching a significant number of students, requires a fairly large computer with time-sharing capability and sufficient terminals to reach the student body. In addition, this system is most effective with the use of graphic CRT display systems. Entire tutorial packages have been developed and are presently in use. The PLATO system (Programmed Logic for Automatic Teaching Operations) combines drill and simulated experiments for the teaching of organic [A-4, A-5] chemistry. A graphical display system at the University of Pittsburgh [A-6] allows students to visualize titration curves in the making even as titration parameters are being varied. The computer involved is an IBM 360-75 located in the University of California at Santa Barbara, almost two thousand miles from its final terminal end-point. There is no doubt that the computer under these circumstances is an effective teacher. The cost, however, can be so large (one estimate places it at about $200 per student) that small colleges can find themselves priced completely out of the market. Again, one of the major difficulties of these systems is that the computer is still geared to the manipulation of numbers rather than word strings, and so the question-and-answer responses still revolve around numerical solutions to specific problems, or to simple yes/no answers.

The breakthrough in small college involvement was the introduction of the small computer, such as the PDP series or the IBM 1130 series. These versatile machines were priced within the budget capability of most any college. Besides limited memory, the major disadvantage of the small computers is that they generally do not have the hardware to support an on-line plotter. However, plotting programs for line printers are about the first item the computer center generally acquires. The real advantage of this type of computer is that it allows hands-on experience for the user. This feeling of controlling the activity of the machine, even for the one-time user, is a big plus for its acceptance. The small computer can act either as a batch-processor or as a single terminal. Although tutorials are possible, time pressure limits their use. In batch, it is possible to develop a sufficient library of "canned" programs to make the student aware that the computer is an ally rather than an adversary. This is most easily accomplished in the areas where laboratory data require some work-up prior to submission as a report. Canned programs should require a minimum of computer skill. The most heavily used program written by this author is a least-squares calculation for a given set of points which presumably are linear. Submitted in batch, the student is required to punch one point per card in entering the x and y coordinates, and to follow the set of data with one blank card. Job control involves two cards which are supplied. For this minimum effort the student is rewarded with an analysis giving slope, intercept, errors, correlations, etc., followed by a neatly plotted graph, bordered on all four sides, accurately scaled, showing both his plotted points and the computer-drawn least-squares line. The plot is a one-page affair, and folded once, can be included with the report. If upon inspection it is obvious to the student that a point is seriously in error, he can remove the one card and resubmit the deck. None of the rules of rejecting data can make as lasting an impression as the simple observation of the computer output. There is every indication that there is a very high motivational incentive for students to use the computer when they find it helpful through their own experience.

Computerized reduction of lab data also provides incentive for upgrading of the laboratory effort. Errorless analysis by the computer places the burden of results on the student with the accompnaying feeling of personal pride in one's endeavors. The criticism that canned programs do not teach the student does have merit, but the simple expedient of having the student perform a sample calculation by hand, if possible, removes this objection. It also reinforces his belief in the accuracy of the computer results. Experiments which might previously have been omitted mainly because of the calculations may now be included with no apprehension. One specific example for which the computer is absolutely essential is in the calculation of molecular coupling constants from nuclear magnetic resonance (NMR) spectra by comparison of the actual spectrum with that simulated by the computer (F-1b]. Another example is the calculation of molecular parameters from infrared spectra [E-21]. Suffice it to say that the computer can be an enormous aid in increasing the depth and comprehension of subject matter in a given time span. Increasingly, experiments in the physical chemistry curriculum involve calculations which can be performed in any reasonably length of time only by the computer. Particularly difficult are experiments in chemical kinetics in which the data may involve numerous chemical species behaving in a nonlinear fashion. Such data are notoriously difficult to reduce.

The collection of programs by DeTar [B-2] and DeMaine and Seawright [B-4] contain programs for the evaluation of kinetic data. DeTar [E-1], Sciano [E-15], and Williams [E-20] have published programs on the handling of kinetic data. Of particular interest is the article by Williams, describing the necessity of proper weighting of data if an accurate least-squares method is to be employed.

Teaching chemical kinetics requires the use of calculus to solve the differential equations represented by the rate laws. For the simple cases these rate laws can be solved analytically. Cases soon arise of consecutive or concurrent reactions where integration of the rate laws is no longer possible. Certain of these equations can be solved by the expedient of assuming a "steady state" concentration of some intermediate. The justification of steady states is generally based on faith--"It works" or "It's the last resort." Computer programs based on simple logic can solve many of these rate laws numerically without any prior assumptions. The Euler method, involving a one-step iteration, is the simplest approach to the solution of differential equations. The first example, the case of consecutive first-order reactions, $A \xrightarrow{k_1} B \xrightarrow{k_2} C$, which does have an analytical solution, can be used to test the algorithm. The idea is to choose a small arbitrary time unit, say 0.02 sec, and to calculate the amount of A used in this period. We then add a small amount to B, calculate the amount of B reacted in this same time, and then increase C. In 50 cycles the time has been incremented by one time unit when the concentrations are all recorded. The initial conditions are: the concentration of A as 100%, B and C both zero, and $k_1$ and $k_2$ having some initial values. The rate expressions are

$$\frac{dA}{dt} = -k_1(A),$$

$$\frac{dB}{dt} = k_1(A) - k_2(B),$$

$$\frac{dC}{dt} = k_2(B).$$

The computer program (Table 1) follows the pattern of the rate equations faithfully. As a check the program calculates the computed value for A and its exact value obtained analytically. The agreement, even with this extremely simple approach is better than 0.1% over the major course of the calculation. PLT3[F-1v] is one of a set of plot routines which can plot three dependent variables versus one independent variable. Figure 1 shows the plotted output resulting from the iterative sequence.

A more difficult example for which a closed analytical solution is not known is the following consecutive bimolecular reaction:

$$A + B \xrightarrow{k_1} X,$$

$$X + B \xrightarrow{k_2} Y.$$

For this reaction the general solution is to assume a steady state of X, i.e., $dX/dT = 0$. This approximation is not required in this algorithm. The rate equations are

$$\frac{dA}{dt} = -k_1(A)(B),$$

$$\frac{dB}{dt} = -k_1(A)(B) - k_2(X)(B),$$

$$\frac{dX}{dt} = k_1(A)(B) - k_2(X)(B),$$

$$\frac{dY}{dt} = k_2(X)(B).$$

The program (Table 2) as written chooses equal amounts of A and B as starting materials and equal rate constants of $1 \times 10^{-5}$. The plotted results as shown in Fig. 2, are quite interesting. One observes that B is the limiting reagent, effectively determining the time span of the reaction, and that indeed the concentration of X reaches a constant value when the reaction is effectively over, clearly supporting the steady state hypothesis. Varying the initial concentrations or the rate constants can give the user an intuitive feeling for the course of the reaction. Analog computers may offer a faster response and may be more applicable than the digital computer in

TABLE 1

FORTRAN Program for Numerical Solution for Kinetics of Consecutive Reaction

| | | |
|---|---|---|
| | REAL K1, K2 | |
| | DIMENSION A(101), B(101), C(101), T(101) | |
| | DATA K1, K2/.10, .05/ | Set rate constants. |
| | DT = .02 | Set time increment. |
| | A(1) = 100. | Initialize time |
| | B(1) = 0. | and |
| | C(1) = 0. | concentrations. |
| | T(1) = 0. | |
| | DO 1 I = 2,101 | Generate 101 time values. |
| | A(I) = A(I-1) | Reset values |
| | B(I) = B(I-1) | for next time |
| | C(I) = C(I-1) | period. |
| | T(I) = T(I-1) + 1. | |
| | DO 2 J = 1,50 | Cycle 50 small time units. |
| | DA = K1 * A(I) * DT | Some A is lost. |
| | A(I) = A(I) - DA | Decrease A. |
| | B(I) = B(I) + DA | Increase B. |
| | DB = K2 * B(I) * DT | Some B is lost. |
| | B(I) = B(I) - DB | Decrease B. |
| | C(I) = C(I) + DB | Increase C. |
| 2 | CONTINUE | |
| | AN = 100. * EXP (-K1 * T (I)) | Analytical value for A. |
| | WRITE 5, A(I), AN | Compare calculated and analytical values. |
| 1 | CONTINUE | |
| | CALL PLT3 (101,T,A,B,C) | Plot results. |
| | CALL EXIT | |
| 5 | FORMAT (2F15.5) | |
| | END | |

this area of kinetics. However, programming the analog device definitely requires some knowledge of electronics. On the other hand, once the algorithm of the digital method is understood, the majority of kinetic problems one faces can be programmed.

```
PAGE   1

// JOB

LOG DRIVE    CART SPEC    CART AVAIL   PHY DRIVE
  0000          2222          2222         0000

V2 M10    ACTUAL   8K   CONFIG   8K

// FOR
*LIST SOURCE PROGRAM
*ONE WORD INTEGERS
*IOCS(CARD,1132PRINTER)
      REAL K1,K2
      DIMENSION A(101),B(101),C(101),T(101)
      DATA K1,K2/.1,.05/
      DT=.02
      A(1)=100.
      B(1)=0.
      C(1)=0.
      T(1)=0.
      DO 1 I=2,101
      A(I)=A(I-1)
      B(I)=B(I-1)
      C(I)=C(I-1)
      T(I)=T(I-1)+1.
      DO 2 J=1,50
      DA= K1*A(I)*DT
      A(I)=A(I)-DA
      B(I)=B(I)+DA
      DB= K2*B(I)*DT
      B(I)=B(I)-DB
      C(I)=C(I)+DB
2     CONTINUE
      AN=100.*EXP(-K1*T(I))
      WRITE(3,5)A(I),AN
1     CONTINUE
      CALL PLT3(101,T,A,B,C)
      CALL EXIT
5     FORMAT(2F15.5)
      END

FEATURES SUPPORTED
 ONE WORD INTEGERS
 IOCS

CORE REQUIREMENTS FOR
 COMMON       0  VARIABLES     826  PROGRAM     228

END OF COMPILATION

// XEQ
```

```
0.36763        0.36978
0.33261        0.33459
0.30092        0.30275
0.27226        0.27394
0.24632        0.24787
0.22286        0.22428
0.20163        0.20294
0.18242        0.18363
0.16504        0.16615
0.14932        0.15034
0.13510        0.13603
0.12223        0.12309
0.11058        0.11137
0.10005        0.10077
0.09052        0.09118
0.08190        0.08251
0.07409        0.07465
0.06704        0.06755
0.06065        0.06112
0.05487        0.05530
0.04964        0.05004
0.04492        0.04528
0.04064        0.04097
0.03677        0.03707
0.03326        0.03354
0.03009        0.03035
0.02723        0.02746
0.02463        0.02485
0.02229        0.02248
0.02016        0.02034
0.01824        0.01841
0.01650        0.01665
0.01493        0.01507
0.01351        0.01363
0.01222        0.01234
0.01106        0.01116
0.01000        0.01010
0.00905        0.00914
0.00819        0.00827
0.00741        0.00748
0.00670        0.00677
0.00606        0.00612
0.00548        0.00554
0.00496        0.00501
0.00449        0.00454
```

```
90.47435       90.48374
81.85603       81.87307
74.05860       74.08183
67.00396       67.03201
60.62137       60.65307
54.84680       54.88117
49.62229       49.65853
44.89542       44.93289
40.61882       40.65697
36.74956       36.78794
33.24888       33.28711
30.08168       30.11943
27.21622       27.25318
24.62369       24.65970
22.27811       22.31302
20.15596       20.18965
18.23595       18.26835
16.49884       16.52989
14.92722       14.95686
13.50531       13.53353
12.21884       12.24565
11.05491       11.08031
10.00185       10.02588
 9.04910        9.07180
 8.18710        8.20850
 7.40723        7.42736
 6.70164        6.72055
 6.06327        6.08100
 5.48570        5.50232
 4.96315        4.97870
 4.49037        4.50492
 4.06262        4.07622
 3.67563        3.68831
 3.32550        3.33732
 3.00873        3.01974
 2.72212        2.73237
 2.46282        2.47235
 2.22822        2.23707
 2.01596        2.02419
 1.82393        1.83156
 1.65019        1.65726
 1.49300        1.49955
 1.35078        1.35685
 1.22210        1.22773
 1.10569        1.11090
 1.00036        1.00518
 0.90507        0.90952
 0.81886        0.82297
 0.74086        0.74465
 0.67028        0.67379
 0.60643        0.60967
 0.54867        0.55165
 0.49640        0.49915
 0.44912        0.45165
 0.40633        0.40867
```

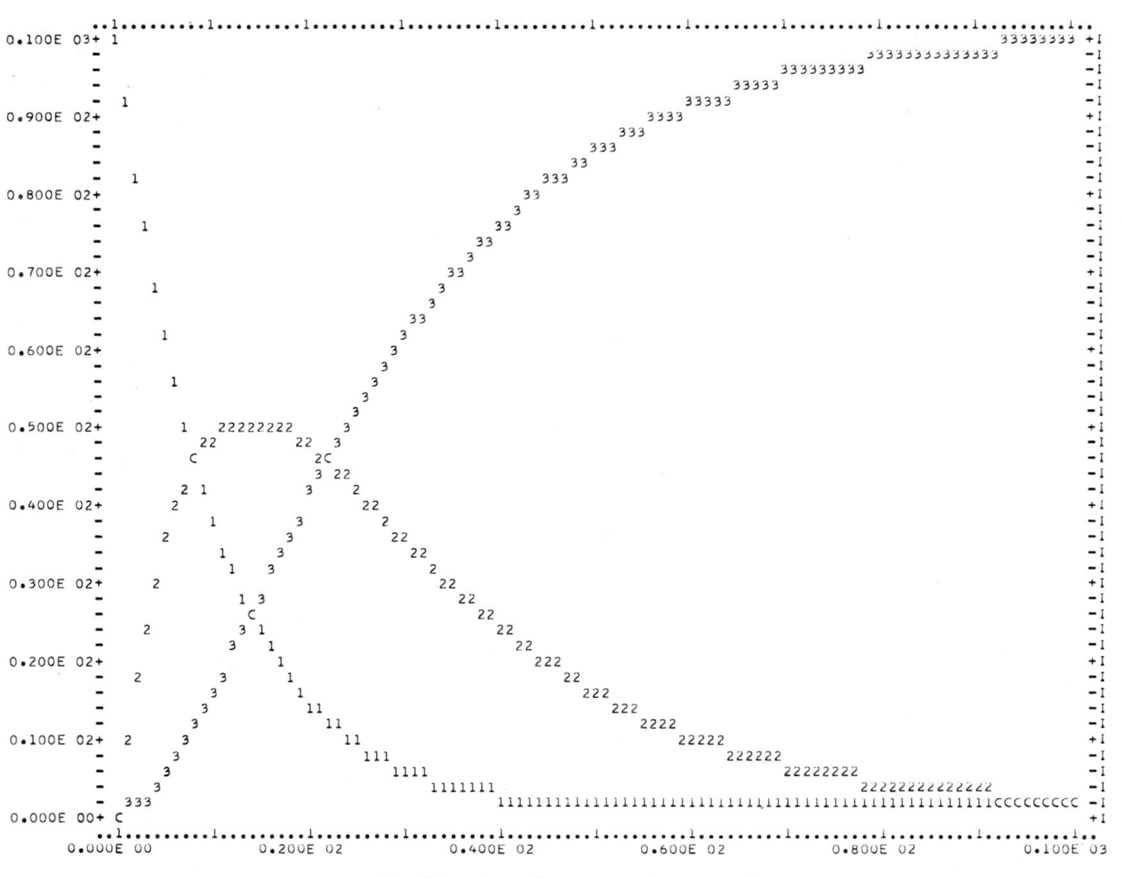


Fig. 1. The kinetics of consecutive reaction.

TABLE 2

FORTRAN Program for Numerical Solution for Consecutive Bimolecular Reaction

```
      REAL K1, K2
      DIMENSION A(101), B(101), X(101),
        Y(101), T(101)
      DATA K1, K2/1.E-5, 1.E-5/            Set rate constants.
      DT = .02                             Set small time unit.
      A(1) = 100.                          Initialize all
      B(1) = 100.                          concentrations and
      X(1) = 0.                                  time.
      Y(1) = 0.
      T(1) = 0.
      DO 1 I = 2,101
      A(I) = A(I-1)                        Reset value for new
      B(I) = B(I-1)                           time unit.
      X(I) - X(I-1)
      T(I) = T(I-1) + 1.
      DO 2 J = 1,50
      DX = K1 * A(I) * B(I) * DT           Some X is made.
      A(I) = A(I) - DX                     Decrease A.
      B(I) = B(I) - DX                     Decrease B.
      X(I) = X(I) + DX                     Increase X.
      DY = K2 * X(I) * B(I) * DT           Some Y is made.
      X(I) = X(I) - DY                     Decrease X.
      B(I) = B(I) - DY                     Decrease B.
      Y(I) = Y(I) + DY                     Increase Y.
2     CONTINUE
1     CONTINUE
      CALL PLT4 (101,T,A,B,X,Y)            Plot results.
      CALL EXIT
      END
```

Many programs, by various authors, are presently available for the reduction of laboratory data usually encountered in the undergraduate physical chemistry laboratory. Corrington [F-2] has published a number of fine programs for the reduction of laboratory data. Occasionally one finds a programmer who has not published in the general literature. Fortnum [F-3] has submitted to this author a substantial list of experiments, details, and programs for the physical chemistry laboratory. This author has written a number of programs as listed in the references [F-1, E-21, E-22]. One of these, [F-1a], has proven quite useful for determining end-points of titrations. The procedure is for the student to punch his observed data, e.g., pH vs milliliters of titrant, on cards and to submit these to the computer. A plot of all the data appears. The student then selects those points which he observes on the vertical section of his sigmoid curve, and resubmits these as a second pass. The computer then returns the first and second derivative of his points, but this time the total scale may involve one or two milliliters at most. Consequently, end-points can be obtained from line printer plots to $\pm$ 0.002 ml, which is more precise than his buret readings, including the computer plotting errors. In practice, while most instructors discuss the improvement in data by use of the second derivative, few actually require the student to use this method on the myriad of titrations generally performed in the course of a semester. In effect, the entire principle of the second derivative is left as an academic exercise. By contrast, the computer teaches the value of the second derivative by experience and by rewarding results.

The closest this author has come to the "ideal" laboratory experiment has been the calculation of the molecular parameters of diatomic molecules [E-21]. Here the student first prepares a gas, usually HBr or DBr, fills a gas infrared cell, and obtains the spectrum on a recording spectrophotometer. The absorption lines are carefully measured using a fine microscope and the data are submitted to the computer for analysis. The point here is that the computer can perform an extremely sophisticated set of calculations which would require months if the student were to do these by hand with a desk calculator. A wealth of information, rotational constants, internuclear distances, etc., is obtained for three states, the excited state, ground state, and zero point energy state. Most texts offer a simplified calculation which is actually some average of the three states effectively discarding much of the information inherently contained in the data. Through a least-squares method, the computer minimizes experimental error, allowing the use of an inexpensive qualitative spectrophotometer.

The lecture part of the chemistry curriculum is generally concerned with the development of the conceptual ideas and the abstract arguments which we call the theories or the laws of nature. Often the non-science oriented student may have serious difficulty grasping the most introductory concepts, as, for example, the units in science, the gram-atom, the mole, the balancing of chemical reactions, etc. Here, the tutorial and drill programs can help alleviate the drain on the instructor's time. As these concepts increase in their mathematical complexities even advanced students may have difficulty assimilating the abstract mathematics into a recognizable physical model. Specifically, quantum chemistry has been moving in a direction away from simple physical models. Most students accept on faith the contention that s atomic orbitals have spherical probability, or that p orbitals have lobes, but to envision the angular character

```
PAGE   1

// JOB

LOG DRIVE   CART SPEC   CART AVAIL  PHY DRIVE
  0000        2222        2222        0000

V2 M10   ACTUAL  8K  CONFIG  8K

// FOR
*ONE WORD INTEGERS
*IOCS(CARD,1132PRINTER)
*LIST SOURCE PROGRAM
      REAL K1,K2
      DIMENSION A(101),B(101),X(101),Y(101),T(101)
      DATA K1,K2/5.E-4,5.E-4/
      DT=.02
      A(1)=100.
      B(1)=100.
      X(1)=0.
      Y(1)=0.
      T(1)=0.
      DO 1 I=2,101
      A(I)=A(I-1)
      B(I)=B(I-1)
      X(I)=X(I-1)
      Y(I)=Y(I-1)
      T(I)=T(I-1)+1.
      DO 2 J=1,50
      DX=K1*A(I)*B(I)*DT
      A(I)=A(I)-DX
      B(I)=B(I)-DX
      X(I)=X(I)+DX
      DY=K2*X(I)*B(I)*DT
      X(I)=X(I)-DY
      B(I)=B(I)-DY
      Y(I)=Y(I)+DY
2     CONTINUE
1     CONTINUE
      CALL PLT4(101,T,A,B,X,Y)
      CALL EXIT
      END

FEATURES SUPPORTED
 ONE WORD INTEGERS
 IOCS

CORE REQUIREMENTS FOR
 COMMON      0  VARIABLES   1024  PROGRAM    230

END OF COMPILATION

// XEQ
```

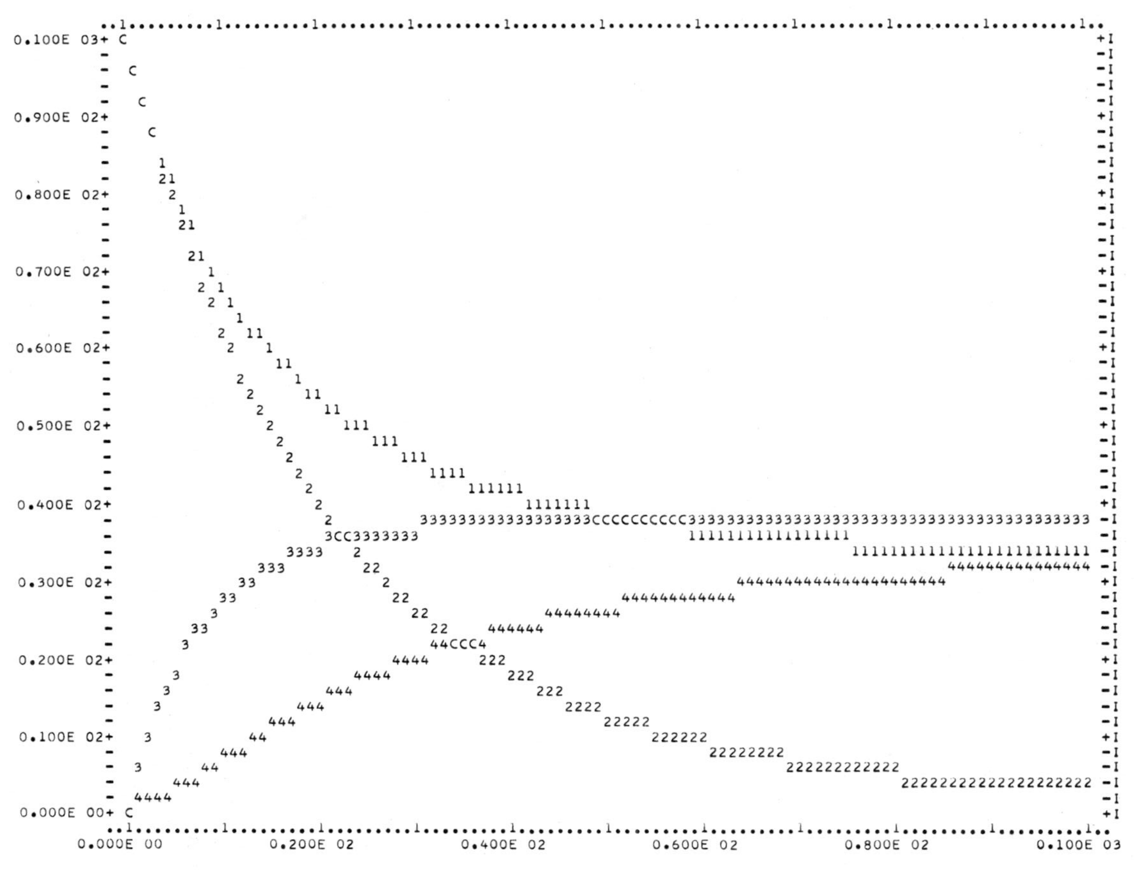


Fig. 2. The kinetics of consecutive bimolecular reactions.

of a hybrid sp orbital certainly requires a mental synthesis uncommon to most scientists, much less students. For that reason, the program HATOM [F-1d] was written. The student can ask the computer in a conversational manner to plot any one of nine atomic orbitals. The computer returns with a scaled contour plot of the orbital, with the geometry clearly evident. The student can then ask the computer to plot a hybrid of any combination of the original nine orbitals. Two parameters are left in the hands of the user, a magnification factor to scale the plot up or down, and a coefficient which allows any variation of mix of the orbitals involved. Thus, the synthesis is performed by the computer with a resulting physical model which is a faithful representation of the mathematics. Another program, HMOLC [F-1e], allows any combination of molecular orbitals to be constructed around two atoms. This particular program constructs nine consecutive frames as a slow-motion sequence, as the two atoms coalesce. The abstract concept of the united atom is crystal clear as the frames unfold. Although finer displays can be demonstrated through artistic films, the educational impact of the computer may be stronger simply because the student is interacting with his observations by initiating and in a sense originating his experiences.

One of the continual amazements is the ingenuity of scientists to simulate events on the computer. One of the most widely used of these is the LAOCOON simulation of NMR spectra [B-2, E-12]. As originally written, this program utilized the entire memory of the large CDC 6400 series computer. By careful reconstruction this program can now be run on a small computer [F-1b] providing another tool for the study of spectroscopy. Recently Carberry [E-18] published a program constructing and plotting the parent peaks of a mass spectrum. Wasson [E-19] introduced a program providing energy level diagrams for electronic spectra of transition metal complexes. In some course in physical or organic chemistry one usually finds some discussion of the empirical Hückel molecular orbital calculations. Wiberg [B-3] has included such a program which is written in Fortran II for a fairly large size machine. Another version of this calculation for 21-atom conjugated systems written in Fortran IV for the small computer has been published by this author [F-1c]. The literature abounds with programs and techniques. The bibliography lists those sources of techniques which have been found most useful in the undergraduate curriculum. Detailed discussion on specific systems and analog devices can be found in the two symposium reports [C-1, C-2].

Many discussions among educators revolve about the question of whether each student should learn to program. The wealth of packaged programs requires only that the student understand the computations involved and be familiar with the data format required for his particular problem. The consensus of opinion observed by this author is clearly in favor of having each student learn a programming language. In a number of schools some laboratory time is devoted to this effort. In nearly all institutions with computer facilities there are formal courses or seminars available to both faculty and students. Often, the problems faced in the elementary chemistry courses could be handled by desk calculators. However, problems do arise in these and later courses for which independently written computer programs are a necessity. These one-time programs should be simple and easily explained with a minimum of programming effort. An example, typical of many small problems taken from a standard text [A-7], asks the student to calculate and plot the potential energy for the interaction of two neon atoms from 2 to 10 Å given the Leonnard-Jones interaction parameters $\epsilon/k = 35.6^\circ K$ and $\sigma = 2.75$ Å. The equation to be used is

$$\frac{U}{R} = \frac{4\epsilon}{k}\left[\left(\frac{\sigma}{r}\right)^{12} - \left(\frac{\sigma}{r}\right)^{6}\right] .$$

The Fortran program listed in Table 3 can be used as a model for many of these short problems.

TABLE 3

FORTRAN Program for Plot of Leonnard-Jones Potential

```
      DIMENSION U (250), R(250)
      DATA EK, SIG/35.6, 2.75/                   Initialize data.
      DA = .05                                   Distance increment.
      A = 2.                                     Initial distance.
      I = 0
2     I = I + 1
      R (I) = A
      B = (SIG/A)**6                             Exponentiation is per-
      U (I) = 4.*EK*B*(B-1.)                     formed only once for time
                                                 saving.
      PRINT 3, R (I), U (I)
3     FORMAT (2E 15.5)
      IF (A - 10.) 4, 5, 5
4     A = A + DA                                 Increment distance.
      GO to 2                                    Repeat.
5     CALL SPLT1 (I, R, U, 2., 10., -40, 60.)    Plot routine with scaling
      CALL EXIT                                  of axes.
      END
```

This program computes and plots the potential at intervals of 0.05Å. Figure 3 shows the result of the computed output. The routine SPLT1 is one of a package of 11 automatic plotting routines [F-1v]. SPLT1 allows the

```
PAGE   1

// JOB

LOG DRIVE   CART SPEC   CART AVAIL  PHY DRIVE
  0000        2222        2222        0000

V2 M10   ACTUAL  8K  CONFIG  8K

// FOR
*LIST SOURCE PROGRAM
*ONE WORD INTEGERS
*IOCS(CARD,TYPEWRITER,KEYBOARD,1132PRINTER,PAPERTAPE,DISK)
      DIMENSION U(250),R(250)
      DATA EK,SIG/35.6,2.75/
      DR=.05
      A=2.
      I=0
5     I=I+1
      R(I)=A
      B=(SIG/A)**6.
      U(I)=4.*EK *B*(B-1.)
      IF(A-10.)3,4,4
3     A=A+DR
      GO TO 5
4     CALL SPLT1(I,R,U,2.,10.,-40.,60.)
      WRITE(3,10)
10    FORMAT(25X,'PLOT OF THE LEONNARD-JONES POTENTIAL FOR THE INTERACTI
     +ON OF TWO NEON ATOMS.')
      CALL EXIT
      END

FEATURES SUPPORTED
 ONE WORD INTEGERS
 IOCS

CORE REQUIREMENTS FOR
 COMMON      0  VARIABLES   1014  PROGRAM    174

END OF COMPILATION

// XEQ
```

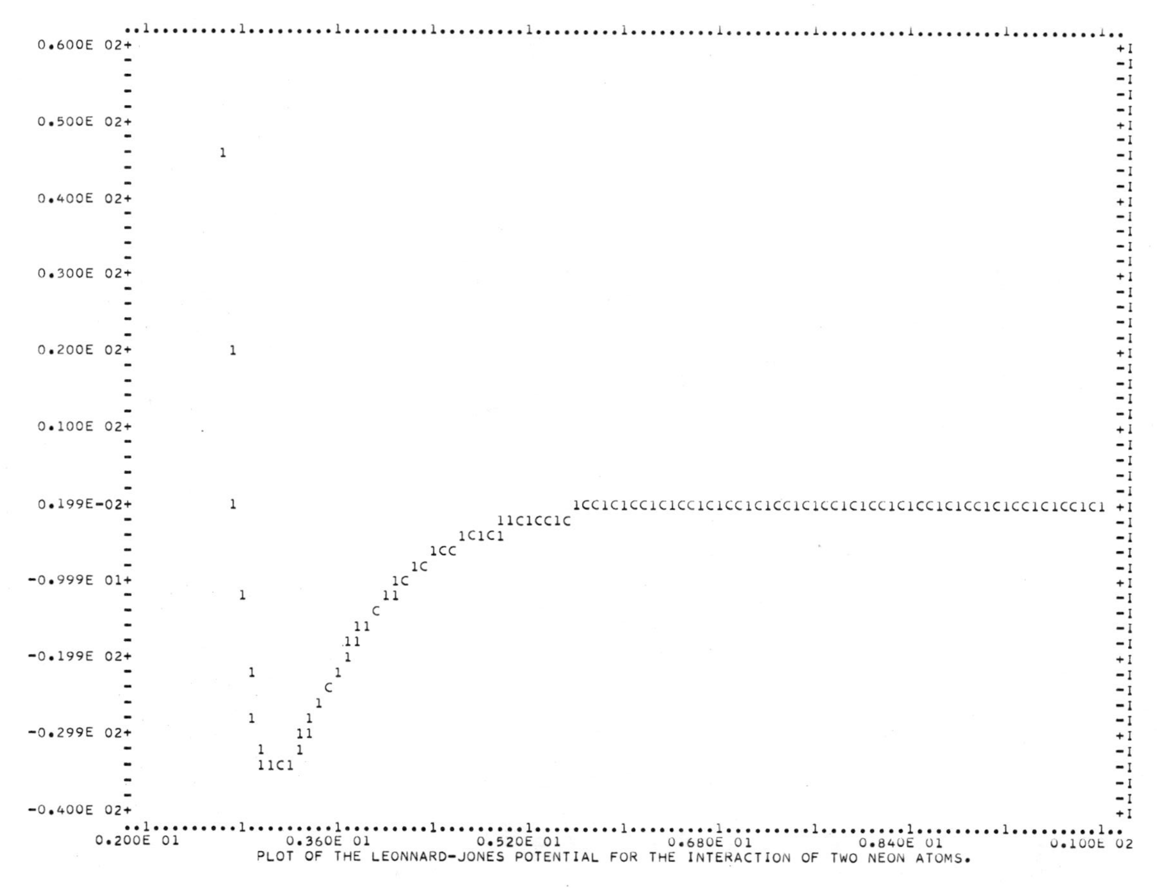


Fig. 3. Plot of the Lennard-Jones 6-12 potential.

user to scale the x and y coordinates. If one has no idea where the points would lie the call CALL PLTI (I, R, U) would automatically scale the axis so that every point would be plotted. The set of plotting programs on the line printer requires no manual for its use and this is the essence of the one-time program.

The bibliography appended here is certainly not complete. It represents those sources which are available to the small as well as the large institution. Particularly relevant are the computer exchanges which are expanding in direction and volume. QCPE, Quantum Chemistry Program Exchange [D-1], now handles many programs which only indirectly have some quantum background. The exchange at Eastern Michigan University [D-2] is relatively new, but has the potential for phenomenal growth. Industrial exchanges also exist [D-3, D-4] but may require some special qualifications for admittance for those outside the organization.

To sum up, the educational system is in the process of dramatic metamorphosis. The future will assuredly see a national network of computers from coast to coast when it will be possible to call a particular computer devoted to a specific area. The duplication of effort to which we are witness must give way eventually to directed effort. Presently, the rewards to instructors for development of pedagogical tools is not commensurate with the rewards given for research and publication. When this inequity is leveled, as it must be, and when the national consciousness awakens to the potentials of the computer in education as a national priority, we will then take a quantum jump into the future.

## III. BIBLIOGRAPHY

### A. General References

1. A. C. Wahl, in Sigma Molecular Orbital Theory (I. Sinanoglu and K. B. Wiberg, eds.), Yale Univ. Press, New Haven, Connecticut, 1969.
2. D. A. Davenport, "The Grim Silence of Facts," J. Chem. Ed., 47, 271 (1970).
3. L. Strong, Earlham College, Richmond, Indiana, private communication.
4. R. C. Grandey, J. Chem. Ed., 48, 791 (1971).
5. S. Smith, Instruction in Chemistry Using PLATO, [Ref. C-2], p. 4-39.
6. K. J. Johnson, Computer-Assisted Instruction at Pitt, [Ref. C-2], p. 4-47
7. F. Daniels and R. Alberty, Physical Chemistry, 3rd ed., Wiley, New York, 1966, p. 27.

## B. Textbooks on Programming in Chemistry


1. T. R. Dickson, The Computer and Chemistry, Freeman, San Francisco, California, 1968.
2. D. F. DeTar, Computer Programs for Chemistry, Vols. I and II, Benjamin, New York, 1968.
3. K. B. Wiberg, Computer Programming for Chemists, Benjamin, New York, 1965.
4. P. A. D. DeMaine and R. D. Seawright, Digital Computer Programs for Physical Chemistry, Vols. I and II, Macmillan, New York, 1963.
5. W. C. Sangren, Digital Computers and Nuclear Reactor Calculations, Wiley, New York, 1960.
6. R. H. Pennington, Introductory Computer Methods and Numerical Analysis, Macmillan, New York, 1965.
7. G. A. Pall, Introduction to Scientific Programming, Appleton-Century-Crofts, New York, 1971.
8. B. Carnahan, H. A. Luther, and J. O. Wilkes, Applied Numerical Methods, Wiley, New York, 1969.


## C. Symposia in Print


1. Proceedings of a Conference on Computers in the Undergraduate Curricula, University of Iowa, Iowa City, Iowa, June 16-18, 1970. Supported by the National Science Foundation.
2. Proceedings of a Conference on Computers in Chemical Education and Research, Northern Illinois University, DeKalb, Illinois, July 19-23, 1971.
3. Computers in Chemical Education, J. Chem. Ed., 47, No. 2, 1970.


## D. Computer Program Exchanges

1. QCPE, Quantum Chemistry Program Exchange, Chemistry Department, Room 204, Indiana University, Bloomington, Indiana.
2. Eastern Michigan University Center for the Exchange of Chemistry Computer Programs, Chemistry Department, Eastern Michigan University, Ypsilanti, Michigan.
3. Perkin-Elmer Program Exchange Library, Digital Applications Laboratory, The Perkin-Elmer Corporation, Norwalk, Connecticut.
4. Computer Programming Library, Packard Instrument Company, Inc., Downer's Grove, Illinois.

## E. Computer Applications to Physical Chemistry


1. D. F. DeTar, Complex reaction kinetic mechanisms, J. Chem. Ed., 44, 191, 193 (1967).

2. M. L. Corrin, Analog representation of first-order reactions, and solution of particle in a box, J. Chem. Ed., 43, 579 (1966).
3. W. C. Hamilton, Computer drawn stereo plots of molecules, J. Chem. Ed., 45, 296 (1968).
4. D. A. Brandreth, Least-squares curve fitting, J. Chem. Ed., 45, 657 (1968).
5. D. T. Cromer, Stereo plots of electron clouds, J. Chem. Ed., 45, 626 (1968).
6. C. S. Ewig, Computer-assisted instruction at the University of Calif. at Santa Barbara, J. Chem. Ed., 47, 97 (1970).
7. E. M. Mortensen and R. J. Penick, Computer animation of molecular vibrations, J. Chem. Ed., 47, 102 (1970).
8. W. Dannhauser, PVT behavior of gases, J. Chem. Ed., 47, 126 (1970).
9. N. C. Craig et al., Computer experiments in enzyme kinetics and wave function contours, J. Chem. Ed., 48, 310 (1971).
10. P. E. Stevenson, Interactive program for hydrogen molecule calculations, J. Chem. Ed., 48, 316 (1971); also published in QCPE (Quantum Chemistry Program Exchange), Indiana University, Bloomington, Indiana.
11. D. L. Peterson and M. E. Fuller, Simulation of the infrared spectra of HCl and DCl, J. Chem. Ed., 48, 314 (1971).
12. L. J. Slazberg, Simple qualitative simulations of NMR spectra and atomic wavefunctions (BASIC), J. Chem. Ed., 48, 449 (1971).
13. E. Hamori, Relaxation kinetics on analog computer, J. Chem. Ed., 48, 39 (1972).
14. L. D. Portigal, 3-dimensional molecular displays, J. Chem. Ed., 48, 790 (1971).
15. J. C. Scaiano, Least-squares curve fitting for kinetic data, J. Chem. Ed., 47, 112 (1970).
16. T. A. Gosink, SCANREC, A program to score tests, analyze results, and record keeping, J. Chem. Ed., 47, 101 (1970).
17. A. A. Zavitas, Calculation of potential energy surfaces, J. Chem. Ed., 48, 761 (1971).
18. E. Carberry, Calculation and plot of mass spectrum isotope peaks, J. Chem. Ed., 48, 729 (1971).
19. J. R. Wasson, Calculation of energy level diagrams and electronic absorbtion spectra of transition metal complexes, J. Chem. Ed., 47, 371 (1970).
20. R. C. Williams and James W. Taylor, Calculation of first-order rate constants, J. Chem. Ed., 47, 129 (1970).
21. M. Bader, A computerized physical chemistry experiment, J. Chem. Ed., 46, 206 (1969).
22. M. Bader, Programs in undergraduate physical chemistry, J. Chem. Ed., 48, 175 (1971).

## F. Programs

### 1. Programs in Chemistry Written by the Author

a. Plot of titration data including first and second derivatives. b. Simulation of high resolution NMR spectra for five spin systems [E-22]. c. Hückel molecular orbital calcuation for 21-atom conjugated systems [E-22]. d. Contour diagrams for hydrogen atom wavefunctions and hybrid orbitals [E-22]. e. Contour diagrams for molecular orbitals [E-22]. f. Calculation of roots of a polynomial. g. Generalized plotting routine for any mathematical expression and any number of its derivatives. h. Demonstration of a Monte Carlo integration. i. Calculations for total charge as measured on a gas coulometer. j. Calculation of molecular parameters of diatomic molecules from the infrared spectra, a computerized experiment [E-21]. k. Calculation of partial molar volumes from raw experimental data. l. Calculation of partial molal volumes from raw experimental data. m. Demonstration of Simpson's rule for integration of the Debye heat capacity function and plot of the total specific heat. n. Calculation of least-squares values for a given set of points and plot of data. o. Dead-time correction for Nuclear Scaler. p. Solution of transcendental equations, can also solve polynomials. q. Least-squares spline fit by orthogonal polynomials. r. Grading of students in analytical chemistry upon submission of triplicate analyses. s. Calculation of buret calibration including air buoyancy corrections. t. Solution of overdetermined simultaneous equations. u. Complete ab-initio calculation of hydrogen molecule from molecular orbital wave function including configuration interaction [based on article by M. J. S. Dewar and J. Kelemen, J. Chem. Ed., 48, 494 (1971)]. v. Package of 11 plotting programs for line printer requiring minimum core and elementary programming.

### 2. Package of Programs and Subroutines for Physical Chemistry by J. A. Corrington, [see Ref. C-1, p. 7.73.]

a. Polynomial regression. b. Correlation of vapor pressure data as a function of temperature. c. Non-linear regression for chemical kinetics. d. Roots of a polynomial. e. Simulation of Van der Waal's PV isotherms. f. Calculation of pH during an acid-base titration per drop of titrant. g. Plot of two independent variables. h. Plot of hydrogen radial wave functions. i. Hydrogenic probability distributions. j. Plot of Slater-type orbitals and Clementi wave functions. k. Plot of two-orbital representation. l. Integration of tabulated values. m. Calculation of bond distance from infrared rotational splitting. n. Calculation and plot of consecutive first-order reactions. o. Calculation of heat capacity for diatomic molecules from partition function. p. Calculation and plot of Leonnard-Jones intermolecular potential. q. Calculation and plot of heat of activation as a function of temperature.

3. Package of Programs for Reduction of Data Obtained in the Physical Chemistry Laboratory, Written by D. Fortnum, Gettysburg College, Gettysburg, Pennsylvania, (unpublished, private communication).

a. Calculation and plot of solid-liquid phase diagram. b. Solubility and heat of solution. c. Vapor pressure of a pure liquid. d. Viscosity of gases by evacuation rate. e. Ratio of heat capacities of gases. f. Partial molal volume. g. Molecular weight by Victor Meyer method. h. Heat of combustion. i. Generalized plotting program.

4. Collection of Programs for Introductory Chemistry Written by C. E. Minnier in BASIC. [C. E. Minnier, J. Chem. Ed., 48, 744 (1971).]

a. Least-squares processing of data. b. Plot of standard error curves. c. Calculation of first- and second-derivatives for titrations. d. Molar volume of hydrogen calculation from reaction of Mg with acid. e. Calculation of moles of water in a hydrate. f. Determination of $SO_3$ in $BaSO_4$. g. Determination of grams of Fe from grams of $Fe_2O_3$. h. Average and standard deviations. i. Plot of error curve given average and standard deviation.

# Section II

# SPECIAL APPLICATIONS

Chapter 4

# COMPUTER-GENERATED REPEATABLE TESTS IN CHEMISTRY

John W. Moore
Department of Chemistry
Eastern Michigan University
Ypsilanti, Michigan

Franklin Prosser
Research Computing Center
Indiana University
Bloomington, Indiana

and

Daniel B. Donovan
Science Department
Corning West High School
Painted Post, New York

## I. INTRODUCTION

Many college chemistry teachers are concerned about underachievement by students in their courses, poor attitudes toward learning and toward science on the part of underachievers, and the many anxieties which arise as a result of poor performance. More and more students avoid chemistry courses as much as possible not because of an innate aversion for the field but from a fear of performing poorly. A variety of proposals to alleviate such negativism have been made [1], but it is often difficult to see how to apply them appropriately within the usual university milieu.

One promising strategy for alleviating such problems is the mastery learning approach [2-7]. This approach asserts that most students can achieve at a level near 80% of criterion for a given course provided that certain conditions having to do with proper specification of objectives, adequate evaluation and remediation, and sufficient allotment of time for mastery are met. Educational research indicates [8-10] that such an approach produces favorable additudinal changes as well as increased achievement. Mastery learning evaluates students more on the basis of the time required for nearly complete mastery of a unit of instruction than the fraction of mastery achieved.

Despite the advantages of the mastery approach, especially for students of average or below-average aptitude, very few college and university instructors have adopted it. While this may be due in part to prejudice against or ignorance of educational theory, in many cases the tremendous logistical problem involved in applying a mastery approach to large numbers of students is at fault. An obvious remedy is to use a computer to handle the large quantities of information involved. Two classes of computer usage may be defined on the basis of whether or not a student has access to on-line interaction with a computer [11].

The computer-generated repeatable test (CGRT) system which we will describe does not involve the on-line interaction of computer-assisted instruction (CAI) and thus is classified as one form of computer-managed [12] or computer-facilitated instruction. It greatly reduces the logistical problems associated with mastery learning without involving its users in the high costs and specialized computer hardware of CAI. Many colleges and universities already have computing systems which can produce CGRTs. The option of adopting CGRT is open to a large number of individual instructors in a variety of disciplines.

## II. USE AND EVALUATION OF COMPUTER-GENERATED REPEATABLE TESTS

The remainder of this chapter is divided into four sections followed by a brief summary. We begin with a general description of the computer programs used to print and score CGRT. The second section describes in detail the use of CGRT in Chemistry C105 (general chemistry for science majors) at Indiana University during the fall and spring semesters of 1970-71. Following this are the results of a study of achievement and attitudes of chemistry students using CGRT. Finally several related systems involving mastery learning in chemistry courses are described and contrasted with CGRT.

### A. General Description of the CGRT System

Large numbers of unique but equivalent tests are generated by a computer program which takes stratified random samples from an item pool and prints out questions and answers in a format appropriate to test-taking and machine- or hand-grading. Under this system students may be examined more frequently and encouraged to keep current. Students have a better opportunity for self-evaluation; they get immediate (within forty-five minutes) feedback since answer keys (including suggestions for remedial study) are provided as soon as the test is over. Most importantly, the exams, because they are unique, may be given repeatably. This allows the student to take a test, discover some of the performance objectives he had not thought of before, study or review to increase his mastery, and take the test again. These exams have real pedagogical value, and much of the trauma associated with testing large classes has been eliminated.

The Computer Generated Repeatable Testing [13, 14] process typically consists of four steps: (1) developing pools of test items, (2) producing tests, (3) administering the tests, and (4) scoring the tests. The second and fourth steps are managed by computer, while the execution of the first and third steps is strongly influenced by the computerized nature of the process.

For each exam, the course instructor develops a pool of items (test questions) which forms the data base from which tests are prepared. This is a rather formidable step. Our experience indicates that one should have at least six to ten items in the pool for each question on an exam to assure adequate variation of the individual tests. An instructor planning to give eight exams of 20 questions each should construct about 1500 individual items for his course. This work, although it has the advantage of being more familiar to most instructors than programming CAI, is every bit as tedious and time-consuming as it sounds. It should be done prior to the first semester in which repeatable testing is to be used in the course. Fortunately, the item pools, once developed, are rather permanent, especially for the basic college undergraduate courses that are the most likely candidates for this computerized testing scheme. Only relatively minor alterations to the item pools are needed to accommodate other instructors, the changes of texts and other changes that may occur in subsequent semesters. Further, textbook publishers often have compendiums of test questions for their popular texts. Reusing test questions semester after semester, or even making the entire item pool available to students, is not a disadvantage under our procedure, and in fact is likely to be distinctly advantageous, since it is merely a specification of behavioral objectives.

Since items will ultimately appear on computer-generated tests, the form of the items must conform to the requirements of present computer printing technology. Normally, items may consist of upper-case letters, numbers, and the usual special characters available on modern high-speed line printers. Diagrams, pictures, and other graphic aids cannot usually be printed directly, although the instructor may easily include these by providing the student with a supplementary sheet of diagrams to accompany the tests. Our experience has been that such "hand-outs" are quite adequate for presentation of structural formulas and other chemical information not readily printed.

If the tests are to be graded manually, technology imposes no limitation on the structure of the answer to an item. The test questions may elicit objective or subjective responses from the student. On the other hand, if the instructor wishes to use mechanical grading techniques, he must provide for a <u>single-character</u> response for each item, because of restrictions imposed by the optical mark sense form readers usually available in universities. While this requirement may appear to be a severe limitation, it in fact allows considerable freedom in the form of objective test items. True-false and multiple choice items call for single character responses. Key-word, fill-in, and other forms resulting in a definite numeric or symbolic answer may easily be reduced to a single character response using the following convention: In such a question the form of the answer is indicated by a series of dots which includes one asterisk. The student will construct his symbolic or numeric answer to the questions, and will record as his response on his mark sense form the single character selected by the position of the asterisk in the string of dots. For example, ..*. means code the third letter or digit of the answer, *... means the first letter or digit, and so forth. Students describe such alphabetically or numerically coded items as being hard but fair. The student cannot answer such an item unless he has mastered the basic concepts and vocabulary. Recall is emphasized; simple recognition is subordinated.

In addition to the question part of an item, which the student sees when he takes a test, each item also has an answer part to allow machine-grading and to provide information to the student after testing. The answer part of an item may contain, in addition to an answer character, any relevant information, such as the full symbolic or numeric answer, textbook page references, and other diagnostic aids for the student. Figure 1 shows the format for both the question and answer portion of a typical chemistry test.

After the instructor has developed a section of his test item pool, he will have it punched onto punch cards or entered into an appropriate editable data file system. Each punch card is coded by means of an eight-digit number which is punched in the last eight columns. The first four digits designate the set to which the card belongs. A set consists of from one to 99 questions of similar difficulty. Usually, questions within a set test for the same behavioral objective. Each question within a set is called an item and is identified by a two-digit number (card columns 77 and 78). Since from three to eight cards are usually required for each item, each card within an item has a unique number in the last two columns.

The cards on which the answer to an item are punched are identified by card numbers beginning with 51 (columns 79 and 80). If machine-grading is to be used the single-character response must appear in card column 33. Comments or remedial information may be punched in columns 35 to 72 of the card. Since as many as 49 answer cards and 50 question cards are permitted for each item, there is no problem with fitting lengthy questions or answers to the punch card format. The item pool used to produce the test in Fig. 1 is shown in Fig. 2.

The individualized tests are generated on a digital computer using a computer program GENERATOR. This program reads the item pool for a particular exam, checks the input data for proper sequencing and correct format, reads information describing the tests to be generated (number of questions per test, etc.), generates and prints the individual test, and punches a small answer summary deck for use in mechanized grading. Each test is individually numbered and has questions on the left part of the line printer page and answers on the right. The item identification numbers for each question appear in the answer part for reference. The instructor will of course separate the answer part from the question part prior to giving a test to the student (see Fig. 1).

The computer program selects items for a test by randomly choosing an item from each set. The order of choosing sets is also randomized. No item is used more than once per test. The digital computer is vital to test production, since the random item selection, formatting, and printing of large numbers of individualized tests is beyond the capacity of nonautomated

```
EXAM NUMBER    1,      FORM NUMBER    1
  SAMPLE CHEMISTRY TEST FROM PROGRAM GENERATOR


QUESTION   1
WHICH OF THE FOLLOWING ARE APPROPRIATE UNITS WITH WHICH TO EXPRESS A
CHANGE IN ENERGY...(READ ALL RESPONSES)    A) CALORIES    B) ERGS
C) LITER-ATMOSPHERES    D) KILOCALORIES    E) BOTH A AND C ARE ENERGY
UNITS    F) ALL OF THESE ARE ENERGY UNITS.

-----------------------------------------------------------------------
QUESTION   2
THE HEAT ADDED TO THE SYSTEM AT CONSTANT ----- IS DENOTED BY THE SYMBOL
DELTA E, OR CHANGE IN ENERGY.
A) ENERGY  B)VOLUME  C)PRESSURE  D)ENTHALPY  F)ENTROPY  G)FREE ENERGY

-----------------------------------------------------------------------
QUESTION   3
BOND ENERGY RESULTS FROM DECREASE IN POTENTIAL ENERGY WHEN TWO ATOMS
COME TOGETHER FROM INFINITE SEPARATION TO FORM A BOND.  THE HIGHER THE
BOND ENERGY THE STRONGER THE BOND.  WHICH OF THE FOLLOWING HAS THE
HIGHEST BOND ENERGY...  A)N2  B)P2  C)(AS)2

-----------------------------------------------------------------------
```

```
EXAM NUMBER   1,    FORM NUMBER    1
  SAMPLE CHEMISTRY TEST FROM PROGRAM GE

QUESTION   1... SET 515, ITEM14


F      A AND D USED FOR HEAT.  C USED FOR
       P-V WORK.  B USED FOR WORK.  SINCE
       DELTA E = Q + W , ANY OF THESE
       UNITS MAY BE USED FOR ENERGY.
-----------------------------------------
QUESTION   2... SET 505, ITEM11

B  SINCE DELTA E = Q + W, AND SINCE
   W = -P (DELTA V), W IS ZERO WHEN
   DELTA V IS ZERO (I.E. CONSTANT
   VOLUME) AND SO DELTA E EQUALS Q.
-----------------------------------------
QUESTION   3... SET 405, ITEM 8


A  SEE P.87 AND FOLLOWING.P2 AND (AS)2
   ARE VERY UNSTABLE.AS THE ATOMS GET
   LARGER, ORBITAL OVERLAP IS LESS
   EFFICIENT AND BONDS ARE WEAKER.
-----------------------------------------
```

```
QUESTION    4
WOULD THE FOLLOWING REACTION BE EXPECTED TO PROCEED SPONTANEOUSLY AT
298 DEGREES KELVIN... (1/2) N2 (GAS) + O2 (GAS) = NO2 (GAS)
A) REACTION WOULD PROCEED (FROM LEFT TO RIGHT AS WRITTEN)
B) REVERSE REACTION WOULD OCCUR (RIGHT TO LEFT AS WRITTEN)
C) SYSTEM IS AT EQUILIBRIUM (REACTION OCCURS AT SAME RATE FORWARDS AND
BACKWARDS)

                                        QUESTION   4... SET 519, ITEM 5

                                        B    CALC. DELTA G OF FORMATION FROM
                                          DATA ON HANDOUT. IF NEGATIVE,
                                          REACTION IS SPONTANEOUS.
------------------------------------------------------------------------------
QUESTION    5
THE ATOMIC THEORY OF MATTER WAS REVIVED DURING THE EIGHTEEN-HUNDREDS
BY JOHN $----.
MARK THE FIRST LETTER OF THE WORD.
                                        QUESTION   5... SET 201, ITEM14

                                        D              P. 166
------------------------------------------------------------------------------
QUESTION    6
IONIZATION ENERGY DOES NOT VARY SMOOTHLY WITH ATOMIC NUMBER SEQUENCE.
PRESENTED BELOW ARE FOUR POSSIBLE WAYS OF PLACING ATOMS IN ORDER OF
INCREASING FIRST IONIZATION ENERGY. (ATOM WITH SMALLEST IONIZATION
ENERGY FIRST).  CHOOSE THE CORRECT WAY.
A) NEON-SODIUM-MAGNESIUM-ALUMINUM    B) SODIUM-NEON-MAGNESIUM-ALUMINUM
C) ALUMINUM-MAGNESIUM-SODIUM-NEON    D) SODIUM-ALUMINUM-MAGNESIUM-NEON.
                                        QUESTION   6... SET 212, ITEM 5

                                        D              P. 190
------------------------------------------------------------------------------
QUESTION    7
USING STANDARD ENTHALPIES OF FORMATION FROM THE HANDOUT SHEET,
CALCULATE DELTA H, THE CHANGE IN ENTHALPY, FOR THE REACTION......
  2 C (DIAMOND)  +  O2 (GAS)  =  2 CO (GAS)
A) 26.41 KCAL    B) -52.82 KCAL    C) -26.86 KCAL    D) -53.72 KCAL
E) -26.41 KCAL    F) 52.82 KCAL    G) 26.86 KCAL    H) NONE OF THESE.
                                        QUESTION   7... SET 517, ITEM 3

                                        D  DELTA H MULT. BY 2 SINCE 2 MOLES CO
                                           FORMED. NOTE THAT DIAMOND IS NOT MOST
                                           STABLE FORM OF C, SO ITS DELTA H
                                           MUST BE SUBTRACTED.
------------------------------------------------------------------------------
END OF TEST        EXAM NUMBER   1,    FORM NUMBER   1
                                        END OF EXAM   1,   FORM NUMBER   1
```

Fig. 1. Sample test as printed by program GENERATOR. Before the test is given to a student the answer part on the right would be separated. This part is given to the student when he hands in his mark sense answer form.

```
REPEATABLE TESTING... INPUT DATA LISTING


THE ATOMIC THEORY OF MATTER WAS REVIVED DURING THE EIGHTEEN-HUNDREDS      2011401
BY JOHN  $----.                                                          2011402
MARK THE FIRST  LETTER OF THE WORD.                                      2011403
                                   D                P. 166               2011451
IONIZATION ENERGY DOES NOT VARY SMOOTHLY WITH ATOMIC NUMBER SEQUENCE.    2120501
PRESENTED BELOW ARE FOUR POSSIBLE WAYS OF PLACING ATOMS IN ORDER OF      2120502
INCREASING FIRST IONIZATION ENERGY. ( ATOM WITH SMALLEST IONIZATION      2120503
ENERGY FIRST ). CHOOSE THE CORRECT WAY.                                  2120504
A) NEON-SODIUM-MAGNESIUM-ALUMINUM    B) SODIUM-NEON-MAGNESIUM-ALUMINUM   2120505
C) ALUMINUM-MAGNESIUM-SODIUM-NEON    D) SODIUM-ALUMINUM-MAGNESIUM-NEON.  2120506
                                   D                P. 190               2120551
BOND ENERGY RESULTS FROM DECREASE IN POTENTIAL ENERGY WHEN TWO ATOMS     4050801
COME TOGETHER FROM INFINITE SEPARATION TO FORM A BOND.  THE HIGHER THE   4050802
BOND ENERGY THE STRONGER THE BOND.  WHICH OF THE FOLLOWING HAS THE       4050803
HIGHEST BOND ENERGY...  A)N2  B)P2  C)(AS)2                              4050804
                                   A  SEE P.87 AND FOLLOWING.P2 AND (AS)2 4050851
                                      ARE VERY UNSTABLE.AS THE ATOMS GET  4050852
                                      LARGER,ORBITAL OVERLAP IS LESS      4050853
                                      EFFICIENT AND BONDS ARE WEAKER.     4050854
THE HEAT ADDED TO THE SYSTEM AT CONSTANT ----- IS DENOTED BY THE SYMBOL  5051101
DELTA E, OR CHANGE IN ENERGY.                                            5051102
A)ENERGY  B)VOLUME  C)PRESSURE    D)ENTHALPY  F)ENTROPY  G)FREE ENERGY   5051103
                                   B  SINCE DELTA E = Q + W, AND SINCE   5051151
                                      W = -P(DELTA V), W IS ZERO WHEN    5051152
                                      DELTA V IS ZERO (I.E. CONSTANT     5051153
                                      VOLUME) AND SO DELTA E EQUALS Q    5051154
WHICH OF THE FOLLOWING ARE APPROPRIATE UNITS WITH WHICH TO EXPRESS A     5151401
CHANGE IN ENERGY...(READ ALL RESPONSES)    A) CALORIES     B) ERGS       5151402
C) LITER-ATMOSPHERES     D) KILOCALORIES     E) BOTH A AND C ARE ENERGY  5151403
UNITS     F) ALL OF THESE ARE ENERGY UNITS.                              5151404
                                   F      A AND D USED FOR HEAT. C USED FOR 5151451
                                          P-V WORK. B USED FOR WORK. SINCE  5151452
                                          DELTA E = Q + W ,ANY OF THESE     5151453
                                          UNITS MAY BE USED FOR ENERGY.     5151454
USING STANDARD ENTHALPIES OF FORMATION FROM THE HANDOUT SHEET,           5170301
CALCULATE DELTA H, THE CHANGE IN ENTHALPY, FOR THE REACTION......        5170302
  2 C (DIAMOND)  +  O2 (GAS)  =  2 CO (GAS)                              5170303
A) 26.41 KCAL    B) -52.82 KCAL    C) -26.86 KCAL    D) -53.72 KCAL      5170304
E) -26.41 KCAL    F) 52.82 KCAL    G) 26.86 KCAL    H) NONE OF THESE.    5170305
                                   D  DELTA H MULT. BY 2 SINCE 2 MOLES CO 5170351
                                      FORMED.NOTE THAT DIAMOND IS NOT MOST 5170352
                                      STABLE FORM OF C,SO ITS DELTA H    5170353
                                      MUST BE SUBTRACTED.                5170354
WOULD THE FOLLOWING REACTION BE EXPECTED TO PROCEED SPONTANEOUSLY AT     5190501
298 DEGREES KELVIN...  (1/2) N2 (GAS)  +  O2 (GAS)  =  NO2 (GAS)         5190502
A) REACTION WOULD PROCEED (FROM LEFT TO RIGHT AS WRITTEN)                5190503
B) REVERSE REACTION WOULD OCCUR (RIGHT TO LEFT AS WRITTEN)               5190504
C) SYSTEM IS AT EQUILIBRIUM (REACTION OCCURS AT SAME RATE FORWARDS AND   5190505
BACKWARDS)                                                               5190506
                                   B       CALC. DELTA G OF FORMATION FROM 5190551
                                      DATA ON HANDOUT.IF NEGATIVE,       5190552
                                      REACTION IS SPONTANEOUS.           5190553
THIS CARD IMMEDIATELY FOLLOWS THE QUESTION AND ANSWER DECK AND MUST HAVE99999999

NUMBER OF DETECTED SEQUENCING ERRORS IN DATA IS    0
```

Fig. 2. Sample item pool used to produce the test in Fig. 1. Due to space limitations only a few items are included in each set.

operations. The instructor may also assign weights (point values) to sets of items, thus allowing him to emphasize particular topics or award points based on the difficulty of items.

The computer time required to generate the tests is very small; the time required to print tests is, however, substantial. Typical times on the

Indiana University CDC 3600 computer system were about four minutes of computer time (of which about 20 seconds are for item selection) to generate 1000 three-page tests, and about three hours of printer time to print them. As shown in Tables 1 and 2, the total cost per test is about five cents, which compares favorably with production of tests by conventional methods.

A student taking a test usually obtains an individualized test (with answer part removed), a mark sense form having 23 possible responses for each question, and a special pencil. He takes a seat in the testing area and immediately enters on his mark sense form his student identification number, the exam number, and his individual test number. The student then marks his answers on his test, and for each question enters the appropriate single-character response on his mark sense form. After completing a test, the student exchanges his mark sense form for the answer part of his individual

TABLE 1

Cost Analysis of CGRT[a]

| Item | Cost |
|---|---|
| Punched cards for item pools (one-time expense)[b] | $ .40 |
| Punch cards for student responses: 1000 cards | 1.00 |
| Printer paper: 3000 sheets | 8.30 |
| Mark sense forms: 1000 forms | 8.80 |
| Keypunching services for item pool punching (one-time expense)[b] | 6.50 |
| Computer charges for test production and grading: about 5 minutes at $200 per hour[c] | 16.50 |
| High speed line printer and controller rental and maintenance: at $1800 per month[d] | 6.00 to 12.00 |
| TOTAL Expenses | $47.70 to 53.70 |

Average cost per test: 4.8¢ to 5.4¢

[a] For 1000 three-page tests.
[b] Prorated over four semesters.
[c] Indiana University CDC 3600 system.
[d] CDC 512 printer system.

TABLE 2

Cost Analysis of Conventionally Prepared Tests[a]

| Item | Cost |
|---|---|
| Paper: 3000 sheets | $ 6.00 |
| Mark sense forms: 1000 forms | 8.80 |
| Punch cards for student responses: | |
| 1000 cards | 1.00 |
| Clerical services at $4.00 per hour[b]: | |
| Typing: 1 1/2 hours | 6.00 |
| Multilithing: 2 hours | 8.00 |
| Collating and stapling: 4 1/2 hours | 18.00 |
| Computer charges for grading: about 1 | |
| minute at $200 per hour[c] | 3.30 |
| TOTAL Expenses | $51.10 |
| Average cost per test: 5.1¢ | |

[a] For 1000 three-page tests.
[b] Estimates supplied by Indiana University, Department of Chemistry.
[c] Indiana University CDC 3600 system.

test. The mark sense form is kept by the proctor for later grading. The student, having the correct answers in hand, can immediately determine his errors and is stimulated to improve his knowledge of weak areas.

The instructor or his assistant will, whenever convenient, have the information on the mark sense forms transformed to punch cards on an optical mark sense form reader. This step is required to obtain a form of input acceptable to the typical academic computing facility; one can bypass this step if optical mark sense form reading equipment is attached directly to his institution's computing equipment.

Scoring of the student responses for an exam is done by computer using program GRADER. Input to this program is the answer summary deck punched by program GENERATOR when the tests were prepared, and the student response cards derived from the mark sense forms. Output of this program is a roster of student IDs and test scores and a punch card deck of the high score for each student for this exam.

The test producing program GENERATOR and the scoring program GRADER are written in FORTRAN. Virtually all academic computing centers have well-maintained FORTRAN compilers that produce a fairly good quality object code. We have several versions of the CGRT programs: well documented ANSI FORTRAN versions designed to run on all commonly available computers and specialized versions of GENERATOR for the CDC 3600 and for the CDC 6600. The specialized versions utilize CDC extensions of ANSI FORTRAN to decrease the execution time dramatically by bypassing the repetitive processing of format statements during test printing. Since GENERATOR is completely output-bound, we anticipate that many potential users of the ANSI FORTRAN version would wish to discuss modification of the program with their systems people to take advantage of local extensions to FORTRAN output facilities.

## B. Administration of CGRT in General Chemistry

A pilot study for CGRT was made by Dr. Jerome Wuller and one of the authors in Chemistry C101 at Indiana University during the fall of 1969. Three different computer-generated forms were used for each hour exam in the course. Mean scores on the three forms administered to about 200 students each during the regular hour exams differed by less than a percentage point, well within the standard deviation of the scores on each form. The tests can, in other words, be made equivalent, even though they are unique, by carefully assigning questions to appropriate sets.

Another result of this preliminary work (which involved no repeatable testing) was the development of some conventions for printing chemical formulas granted the limitations of the usual line printer. Appropriate use of parentheses and bond lines often permitted intelligible (though sometimes unusual) formulas to be printed. Some examples are: ethanol, CH3-CH2-OH; manganese dioxide, (MN)O2; nickel carbonyl, (NI) (C=O)4; and glycine, NH2-CH2-COOH. Symbols having two letters are enclosed in parentheses. Diatomic ligands (such as -CO) have bond lines (- or = but not ≡ are available) inserted. Beginning students seem to have less trouble adapting to such a convention than do their instructors.

More complicated structural formulas were presented by means of "hand-outs" which were numerically coded to correspond with certain sets of questions. Each hand-out could contain 20 to 30 structural formulas. Hand-outs have the disadvantage that if a student is permitted to remove one from the testing area and then return with it later, a good deal of useful information (in addition to the formulas) may have been added by one of his more knowledgable compatriots.

The computer programs and other aspects of CGRT described in the preceding section were first used in a large-scale chemistry course during the fall semester of 1970 at Indiana University. All 700 students enrolled in Chemistry C105, a course for science majors, were examined by means of repeatable tests on seven different occasions during the semester. In each case exams were made available at the beginning of a two-week period. After a deadline at the end of this period students were not permitted to take additional tests without a medical excuse. Lectures on the material included in the test usually were completed one or two class periods prior to the deadline. Students were allowed up to three tests on each exam, with only the highest score recorded for the final grade.

Each student reported to an examination room with 100 seats to take the tests, and was permitted to take no more than one test per day so that the large number of students in the class could receive equitable treatment. The exam room was open 20 hours per week on the average, but more time (seven hours per day) was allotted just before a deadline. Each student had been instructed to bring a slide rule, a No. 2 lead pencil, his identification card, and a molecular model kit (if desired). Upon entering the examination room, the student was asked to show his ID card and was not given a test if he failed to produce it. The proctors handed each properly identified student a test (minus the matching answer key), hand-outs and periodic table, and a mark sense form for recording answers. The student was cautioned to mark his ID number carefully on the mark sense form, since improper identification would result in the loss of the test score when the computer matched scores with ID numbers. The exam and form number (unique for each test) also had to be indicated properly on the mark sense form so that the responses could be computer-matched with the correct answers. Two proctors supervised the testing and were permitted to help students with the general interpretation of questions. The removal of hand-outs from the test room was forbidden and a student found to have a hand-out with extraneous information on it received a zero for that test. Upon completion of a test the student submitted the mark sense form to a proctor and received the answer key to his particular test. The student was permitted to take the test and answer key from the test room and use this directive feedback immediately. The student was encouraged to seek help in answering unresolved questions from the other students, discussion section teaching assistants, or the instructor himself. For those students who did not take all three tests before the deadline date, a last test was given in a large lecture room on that date under the same procedure employed in the small test room, but with many more proctors.

Another feature of the testing system employed at Indiana University was the provision of complaint sheets. If a student felt that a question was answered incorrectly on the key or was ambiguous, he had the option of submitting a complaint sheet which identified the question by exam, test, and item number. Each complaint was reviewed by the course instructor, commented upon and regraded, and returned to the student. If a score adjustment was necessary, it was easily handled by a computer program, which also provided records of student performance for course instructors, discussion section leaders, and laboratory assistants.

The examination scores, identified only by ID number, were posted within two days after an exam was completed, and students were encouraged to check their scores to see if there were any discrepancies between the posted score and the score the student figured he had obtained. This comparison was possible because most students marked their answers on the test form as well as the mark sense form.

The second semester class of approximately 400 students, designated as the experimental group, did not take repeatable tests for the first half of the semester but were supplied with a set of answered problems as an aid to study at least a week prior to each scheduled exam. A single unit examination was administered in a large lecture room, under the usual circumstances for large lecture classes. Computer-printed test forms were used and answer keys were provided in a nearby room immediately following the test. Complaint sheets were also available. Since these exams were given during the heavily scheduled morning hours, instead of being taken at the student's convenience, immediate protracted discussion of the exam usually was not possible. For the latter half of the semester, the second semester class used repeatable testing conditions like those described for the first semester class.

The computer-generated repeatable testing scheme used in Chemistry C105 was imposed on an already existing course structure. It is of interest to consider what changes in that structure suggested themselves as the experiment proceeded.

The classical course structure consisted of three 45-minute lectures, a 45-minute discussion section, and a 165-minute laboratory section per week. Taking repeatable exams added one or two hours to this schedule every two weeks since a majority of students availed themselves of the opportunity for three tests on each exam. In order to provide adequate time for students to take tests in the exam room additional assistants had to be assigned as proctors. At the same time, however, attendance (which was optional) at the discussion sections dropped and because all quizzes were machine-graded the workload for discussion assistants dropped considerably.

Student motivation reached a peak for each individual immediately following taking a test. So much so, in fact, that one of the authors recalls being "cornered" for 30 to 60 minutes on several occasions by students clamoring to learn chemistry! At least to learn enough to be able to answer the types of questions on the tests.

As a result of the factors mentioned in the preceding paragraphs, we concluded that a "double testing room" would be more appropriate than the single room we used. Tests would be administered in one room following which students could consult a teaching assistant or faculty member in a nearby room to clear up any lack of understanding revealed by the test. To staff the second room the normal system of regularly scheduled discussion sections would be scrapped and assistants assigned to several hours per week in the testing rooms instead. Such an arrangement has the additional advantage that less experienced assistants may be assigned as proctors while the better teachers are assigned tutoring duties.

Unfortunately these changes could not be effected during the 1970-1971 academic year because they would have increased the number of confounding variables encountered in the experiments described in the next section. It remains for the future to try them.

## C. Evaluation of CGRT

As previously noted in the introduction, the proponents of mastery learning schemes predict increased student achievement and more positive student attidues. An attempt was made to evaluate each of these claims during the spring semester of 1971.

The course outline, instructor, textbook, and grading scheme (except for upward-curving of the nonrepeatable unit exam scores) were identical for C105 during both fall and spring semesters. During the fall repeatable tests were given for all units, but in the spring the first three exams were nonrepeatable, while the last three were repeatable.

To test student achievement a nonrepeatable final examination was given during the regularly scheduled two-hour exam period. The same forms were used both in the fall and the spring. In order to insure the security of the final examinations each student was required to submit his question sheet and answer form individually when he had completed the exam and proof of identity was required. Exams were given in small rooms to small groups of students and each proctor made certain that every student in his room returned an exam.

The final examination was divided into two parts. Part I covered material presented during the first half of the course (when spring semester students had nonrepeatable tests). Part II covered material presented during the second half of the course (both fall and spring semester students were taking repeatable tests). Table 3 summarizes student performance on both parts of the final exam.

The only case where a significant difference in achievement occurs is on Part I of the final exam where students who took repeatable exams averaged three percentage points higher than those taking nonrepeatable exams. However, the spring semester students also scored significantly lower in the mathematics portion of the scholastic aptitude test, a measure which correlates with success in general chemistry [15, 16]. Thus, the difference in achievement scores may reflect this difference in mathematics aptitude. On the other hand, despite the aptitude difference, there was no significant difference in scores on Part II of the final exam (when both groups studied with repeatable exams). The implication is that repeatable testing makes learning less dependent on specific aptitudes. That is, poorer students may be brought to a higher level of mastery if proper strategies are employed. More details regarding analysis of the achievement data may be obtained from Ref. [17].

In order to evaluate changes in student attitudes a 24-item measure was constructed in a manner similar to that of Schwirian [18], using a five-point Likert-type scale [19]. This survey was administered to the spring semester class on four equally spaced occasions. Two of these administrations were performed when the students were using answered sample problems and single

TABLE 3

Student Achievement[a]

| Student group | Mean scores | | | | |
|---|---|---|---|---|---|
| | Part I | Part II | Final total | SAT-V | SAT-M |
| Fall semester | 70.5 | 68.6 | 139.1 | 525 | 578 |
| Spring semester | 67.5 | 66.5 | 135.0 | 516 | 554 |
| F ratio[b] | 5.175* | 1.644 | 3.531 | 1.085 | 8.144* |

[a] Details of the evaluation of student achievement are found in the text.
[b] All cases where the F ratio is large enough to indicate a significant difference in scores (at the 0.05 level) are marked by an asterisk (*).

examinations; the other two administrations were performed when the subjects were under the treatment employing computer-generated equivalent repeatable tests. On each occasion, a uniform procedure was used in an attempt to place the students in such a frame of mind that they would seriously consider their responses to the questions, and the fact that their responses would remain anonymous was emphasized. A conscious effort also was made to address the subjects in a similar manner on each administration of the measure so that no bias would be introduced into any set of scores.

The results of the attitude survey are summarized in Table 4. The results of the two administrations under single-unit exams have been averaged, as have those for the two surveys under repeatable exam conditions. Some of the statements about which student attitudes were surveyed were worded slightly differently on the first two forms to take account of the different treatment of the sample group. For instance, statement number 10 in Table 4 was, "Providing sample examination questions should make your achievement less dependent on the quality of instruction" when nonrepeatable tests were given. On the two surveys during repeatable testing this was changed to, "Repeatable examinations should make your achievement less dependent on the quality of instruction." A complete list of statements for the attitude survey is found in Ref. [17], as is a more complete analysis of the results.

In general student attitudes improved when the change from nonrepeatable to repeatable exams was made. Except for statements 2, 4, and 14, what we considered to be positive attitudes were improved. Changes in attitude toward statements 2 and 14 may reflect the fact that some questions which had not been discussed in class were deliberately added to the repeatable tests. That is, the exams were being used as a teaching device.

The greatest improvement in attitude was toward statements 7, 12, and 15. Students felt that they were learning by taking exams and they felt much less pressure to make good grades. Moreover, although students were undecided about whether answered sample problems made them feel less compulsion to cheat, they clearly agreed that repeatable exams removed some pressure towards cheating.

The population to which the results of this study are generalizable is restricted by purely statistical considerations to all those students who took chemistry C105 during the 1970-1971 academic year. However, it is interesting to note that male students at Indiana University had a mean SAT-V score of 461 and a mean SAT-M score of 514. These values for all male high school seniors tested by Educational Testing Service at the same time as the Indiana University students were 457 and 506, respectively.

TABLE 4

Analysis of Attitude Survey[a]

| Statement | Mean scores[b] | | F |
|---|---|---|---|
| | Non-CGRT | CGRT | Ratio[c] |
| 1. The course is more than an average intellectual challenge | 3.22 | 3.49 | 18.57* |
| 2. One does not have to do too much work for the credit earned | 4.21 | 3.96 | 13.79* |
| 3. A student learns more if he is required to take many exams | 3.76 | 4.09 | 23.08* |
| 4. A student's attitude toward chemistry should be more positive after taking C105 | 3.69 | 3.55 | 6.16* |
| 5. The lecture is the aspect of the course most relevant to learning | 3.51 | 3.42 | 1.65* |
| 6. The laboratory is that aspect of the course least relevant to learning | 3.87 | 4.16 | 23.86* |
| 7. A student learns while in the process of taking examinations | 3.81 | 4.25 | 49.08* |
| 8. You learn more if you keep the exams and use the answer key as a guide to remedial study | 4.25 | 4.47 | 21.07* |
| 9. Discussing one another's exams increases knowledge of course material | 4.01 | 4.27 | 24.55* |
| 10. Achievement is less dependent on the quality of instruction | 2.84 | 3.23 | 29.90* |
| 11. The grade you are receiving is an accurate indication of your comprehension of course material | 2.81 | 3.19 | 22.13* |
| 12. You can take an examination without feeling pressure to make a good grade | 2.67 | 3.29 | 53.73* |
| 13. You have to study harder for exams | 3.59 | 3.82 | 11.95* |

TABLE 4 (continued)

| Statement | Mean scores[b] | | F |
| --- | --- | --- | --- |
| | Non-CGRT | CGRT | Ratio[c] |
| 14. The exams do not contain too many questions which rely on material not presented in C105 | 3.33 | 3.12 | 10.27* |
| 15. A student should feel less compulsion to cheat on an exam | 3.16 | 3.59 | 46.75* |

[a] The administration of the attitude survey is described in the test.
[b] Student agreement with the statement is rated on a scale of 1 = disagree strongly, 2 = disagree, 3 = undecided, 4 = agree, 5 = agree strongly.
[c] All cases where the F ratio is large enough to indicate a significant change of attitude (at the 0.05 level) are marked by an asterisk.

Female students at Indiana University had a mean SAT-V score of 462 and a mean SAT-M score of 474. The values for all female high school seniors tested at the same time as their peers who were now Indiana University freshmen were 458 and 461, respectively. The similarity of the means of Indiana University freshmen to the national means leads one to speculate that the achievement results noted might be generalized to a larger population.

### D. Other Applications of Computers to Mastery Learning

We have purposely described the CGRT system in rather general terms in Section IIA, because the computer programs GENERATOR and GRADER may be used in other ways than the specific application to Chemistry C105 described in Section IIB. In this section we note briefly some other applications of computers to mastery learning which are closely related to CGRT. The interested reader may wish to consult directly with those whose work appears most closely related to a specific application.

Program GENERATOR may be used to produce individualized homework assignments which can be scored using GRADER. Yaney [20] has reported a similar system using programs which he devised. This approach has been used successfully at Eastern Michigan University. A related approach is the Classroom Teacher Support System developed by IBM and the Los Angeles City Unified School District [21]. Approximately 100 teachers used this latter system to obtain quiz or homework questions in several subject matter areas, but chemistry has not as yet been included.

The so-called "Keller Plan" [6, 22, 23] is another area where GENERATOR may be used profitably. The Keller Plan is a self-paced, student-tutored, mastery-oriented learning system. It requires that students pass (at mastery level and in proper order) unit tests for as many as 30 short units of subject matter per semester. Tests are graded by student tutors. Professor Hans Brintzinger [24] of the University of Michigan has adapted GENERATOR to produce tests on a teletype terminal at student demand. Since students will be taking a given unit test over a period of several weeks or more, the ability of the computer to generate a large number of unique, equivalent tests is extremely useful.

Another logical extension of computerized testing is to have the computer write the questions itself, rather than simply selecting questions from an item pool. One program of this type has been written by one of the authors, but a more extensive application has been reported by Miller et al. [25]. The programs used to generate questions also printed tests on an IBM Selectric terminal. A special "Chemical" element (ball) was used so that most of the symbols (lower-case letters, sub- and superscripts, etc.) needed by chemists were available. While there was no major problem involved in using the limited character set available at Indiana University, this approach has obvious advantages.

Another means of obtaining greater flexibility in printing is to obtain a special print train for a line printer. Jameson [26] at the University of Illinois at Chicago Circle has used this approach so that lower-case letters and sub- and superscripts may be printed. This probably represents the least expensive means of producing large numbers of tests which are printed in a format most familiar to chemists. It should be pointed out, however, that students who are not familiar with typographical conventions used by chemists seem to have less trouble adapting to the limitations of the standard line printer than do their teachers.

## III. SUMMARY

Computers have often been used as a tool to alleviate problems caused by the tremendous numbers of students enrolled in introductory courses at large universities [26-28]. Most of these applications, however, have simply involved automation of teaching procedures which were already in common use. The computer-generated repeatable test system differs significantly from such approaches because it involves a mastery learning approach and is strongly student-oriented rather than subject matter oriented.

Performance objectives [29-31] are specified by CGRT in the clearest form possible--as examination questions. While students may easily ignore or fail to understand fully a list of objectives that is posted or handed out, few are so lacking in self-respect as to ignore poor performance on a test or quiz. Once objectives have been specified, CGRT allows a student to evaluate his understanding of subject matter and provides almost immediate feedback, including suggestions for remedial work and additional study. An additional advantage of CGRT and other mastery learning approaches is that a "curve" is no longer needed. An absolute scale of achievement is provided against which the student can match himself; grades become more a measure of how much time and effort a student is willing to spend to reach a given level of mastery than of what fraction of mastery has been reached by a given deadline.

Students taking CGRT were amazed to find that their instructors were not worried about giving "too many A's," provided that everyone performed at an adequate level. Once they got used to this idea it was each individual against the computer. Teachers and other students were viewed as sources of help and information rather than as evaluators (who might detect lack of knowledge if too many questions were asked of them) and competitors, respectively.

Another aspect of the "curve" which became evident during the spring semester experiment was the difference in a student's perception of an "A" obtained as a result of curving a single unit exam and an "A" obtained through "honest effort" on the second or third repeatable test. Students seemed to place greater value on attaining through their own efforts a goal set ahead of time on an absolute scale. Grading on a curve appears to students to be an admission of failure on the part of the teacher, who is willing to concede a few high grades even though his standards have not been met.

An advantage of CGRT which is not immediately evident from our previous discussion is its effect on the teachers who use it. In fact, the same thing would appear to be true of other applications of computers in chemical education [32]. Writing good test questions which can serve as an adequate specification of objectives as well as providing useful remedial feedback is not an easy task. The CGRT system (especially the "complaint sheets" mentioned in Section IIB) provides for feedback to the teacher from students as well as offering a motive for improving test questions because they may be used over and over again. All of the elements (motivation, drill and practice, remedial feedback) which can help the teacher learn more about testing are provided and so, in fact, the teacher does learn about testing.

There is one drawback of CGRT and other systems which are based on behavioral objectives which cannot be ignored. The rules [30] for writing objectives (or test questions) are such that it is much easier to test for some things than for others. The reason is that a particular behavior on the part of the students is necessary in order to state that the objective is met. In trying to teach equilibrium in aqueous solutions of weak acids or bases, an objective might state that given the analytical concentration of a weak acid or base as well as the $K_a$ or $K_b$ a student will be able to calculate the correct pH nine times out of ten. But suppose one wished to determine whether a student has an appreciation and understanding of the wide-ranging applicability of the Second Law of Thermodynamics. Writing a well-stated behavioral objective (or test question) is far more difficult and therefore less likely to be done. As a result, one must beware of devoting too much time to trivia which are easily tested for and skimping on general concepts of broad applicability.

Of course, the caveat embodied in the previous paragraph applies to any teaching situation, not just repeatable testing. In fact, the repeatable tests allow one to experiment with questions which require application of principles to specific problems, because a student is not greatly penalized if he does not reason correctly on his first try. Nevertheless, such considerations are often ignored [31] in chemistry courses, and they should not be.

In addition to the statistics presented in Section IIC, the subjective feelings of those who have been involved in repeatable testing may be of interest. Almost everyone involved has become quite enthusiastic about the system, despite the amount of work necessary to get it started. Faculty members from Indiana University campuses in other locations than Bloomington as well as other universities have adopted the technique. In fact, the student grapevine within the state of Indiana has spread the word and considerable demand has arisen for adoption of CGRT. Other disciplines have used the same programs successfully, too. The reasons for all these favorable reactions are implicit, we think, in any mastery learning approach. Students do not mind putting additional effort into a course if there appears to be a reasonable opportunity to improve their performance. Computer-generated repeatable tests are one means of providing such an opportunity; it is because of this that they are a very successful teaching device.

## REFERENCES

1. B. Z. Shakhashiri, Chem. and Eng. News, July, 12, 1971, p. 41.
2. B. S. Bloom, Mastery Learning: Theory and Practice, Holt, Rinehart and Winston, New York, 1971, pp. 13-28.
3. J. H. Block, The Effects of Various Levels of Performance on Selected Cognitive, Affective and Time Variables, Doctoral Dissertation, University of Chicago, 1970.

4. R. F. Biehler, Educational Psychologist, 7(3), 7-9 (1970).
5. K. M. Collins, An Investigation of the Variables in Bloom's Mastery Learning Model for the Teaching of High School Mathematics, Doctoral Dissertation, Purdue University, 1971.
6. F. S. Keller, J. Appl. Behavior Anal., 1, 79-89(1968).
7. M. D. Merrill, K. Barton, and L. E. Wood, J. Ed. Psych., 61, 102-109 (1970).
8. W. B. Brookover, T. Shailer, and A. Patterson, Sociology of Ed., 37, 171-178(1964).
9. N. T. Feather, J. Personality and Social Psych., 3, 287-298(1966).
10. C. C. Modu, Affective Consequences of Cognitive Changes, Doctoral Dissertation, Univesity of Chicago, 1969.
11. P. Suppes and M. Morningstar, Science, 166, 343-350(1969).
12. W. C. Cooley and R. Glaser, Science, 166, 574-582(1969).
13. F. Prosser and D. D. Jensen, AFIPS-Conference Proceedings, 38, 295-301(1971).
14. F. Prosser and J. W. Moore, Proceedings of the Conference on Computers in Chemical Education and Research, Northern Illinois University, DeKalb, July 19-23, 1971, p. 9-26.
15. N. A. Sieveking and J. C. Savitsky, J. Res. Sc. Teaching, 6, 374-376(1969).
16. B. O. Hendricks, C. L. Koelsche, and J. C. Bledsoe, J. Res. Sci. Teaching, 1, 81-84(1963).
17. D. B. Donovan, Computer-Generated Repeatable Examinations and College Chemistry Student Achievement and Attitude, Doctoral Dissertation, Indiana University, 1972.
18. P. M. Schwirian, Sci. Ed., 52, 172-179(1968).
19. R. Likert, Arch. Psych., 22(140), 1-55(1932).
20. N. D. Yaney, J. Chem. Ed., 48, 276(1971).
21. G. Lippey, F. Toggenburger, and C. D. Brown, Association for Educational Data Systems Journal, March 1971, pp. 75-84.
22. Ben A. Green, Jr., Am. J. Phys., 39(7), 764-775(1971).
23. B. A. Green, Jr., J. Coll. Sci. Teaching, 1(1), 50-52(1971).
24. H. A. Brintzinger, personal communication.
25. W. Miller, C. Howard, M. M. Looney, and C. A. Lewis, Computerized Testing for Self-Paced Instruction, Abstract No. 154, 27th Southwest Regional A. C. S. Meeting, San Antonio, December, 1971.
26. C. C. Hinckley and J. J. Lagowski, J. Chem. Ed., 43(11), 575-578 (1966).
27. K. M. Wellman, J. Chem. Ed., 47(2), 142(1970).
28. C. B. Leonard, Jr., J. Chem. Ed., 47(2), 149-151(1970).
29. R. W. Burns, Ed. Tech., 8(18), Sept. 30, 1968.
30. R. F. Mager, Preparing Instructional Objectives, Fearon Publishers, Palo Alto, California, 1962.

31. S. Markle, The use and abuse of behavioral objectives, in Chemical Education for Underprepared Students (R. I. Walter, ed.), Stipes Publishing, Champaign, Illinois, 1971.
32. F. D. Tabbutt, Chem. and Eng. News, Jan. 19, 1970, pp. 44-57.

Chapter 5

# USE OF A TIME-SHARED COMPUTER FOR LIVE CLASSROOM DEMONSTRATIONS OF CHEMICAL PRINCIPLES

Philip E. Stevenson

Department of Chemistry
Worcester Polytechnic Institute
Worcester, Massachusetts



## I. THE COMPUTER IN THE CLASSROOM

The time has arrived for the electronic computer to assume its place in the classroom as a teaching tool alongside such aids as the sound recording, motion picture films, slides, live demonstrations, wall charts, and the blackboard. An instructor who has interactive access to a computer from his classroom can easily demonstrate numerical results of calculations which are simply impossible to generate at a moment's notice in any other way. He has, of course, obtained or written his program ahead of time and tested it thoroughly, just as he would prepare and test a live demonstration ahead of time.

The computer can be used as a tutor, with the student receiving his lessons directly from the computer and proceeding at his own pace without the direct presence of the instructor. Much work has been done in this area, as described elsewhere in this book, but the requirements of a large machine, very sophisticated programming, and a large number of terminals for student access places tutorial use of the computer beyond the financial capabilities of most colleges. Routine student access to computers for problem-solving has become very common in college education. Normally student-written problem-solving programs are run in a batch environment, but many colleges have followed the lead of Dartmouth in providing time-sharing to their students. However, even this kind of educational use of computers can be more expensive than many colleges can afford. In contrast, the use of demonstration programs in the classroom can bring the computer into effective use in education at a relatively modest cost. All the hardware that is necessary is a single computer terminal in the classroom, tied to a remotely located computer. Alternatively, the class can be brought to the location of a small computer, such as an IBM 1130.

The use of interactive computer programs in the classroom adds a new dimension to the teaching of mathematically complicated topics, and many such topics exist in the chemistry curriculum. These topics range from ionic equilibrium in freshman chemistry to advanced areas in quantum mechanics and statistical mechanics. While teaching these topics, the instructor equipped only with the conventional blackboard and chalk simply cannot demonstrate quantitative application of these formulas, except perhaps by reference to previously calculated quantities, or by the use of hand-waving approximations (as in ionic equilibria), or by the use of very simple idealized cases (such as the Ideal Gas Law). However, if he has a computer terminal at his side and an already written and stored program which can do the computations he wishes to demonstrate, his discussions of such topics can assume a quantitative shape not easily attained any other way. For example, a discussion of gas laws can be augmented by use of a program which calculated van der Waals isotherms for arbitrary temperatures and substances. The instructor can generate numerical results, not only for cases which he prepared ahead of class, but also for cases which occur to him in class and, more importantly, cases suggested by students. This ability to give numerical answers to student questions is probably the most important function of the interactive computer in the classroom.

## II. PROGRAMMING THE COMPUTER FOR CLASSROOM USE

The computer most often enters the classroom in the form of a remote teletype terminal. The instructor must therefore use keyboard input to "talk" to the computer. Since he is simultaneously attempting to hold the attention of his class, the amount of input should be held to a minimum. The computer "answers" by typing the results of its computations on the teletype paper, generally at the slow rate of ten characters per second or seven seconds per full line of typing. Minimization of the amount of output is therefore also advisable. The ideal demonstration program thus processes one input parameter and yields a single number for a result.

Display of computed results to the class can be handled several ways. A small enough class can simply "gather around" and watch the terminal type. In larger classes where gathering around becomes impractical the instructor could read off the results to the class or copy them onto the board. More efficient, however, would be the use of overhead projection or closed circuit television. (Indeed, an electronics package has recently been developed which provides a direct interface between a keyboard terminal and a TV monitor, allowing the elimination of the TV camera.) The normal teletype line is 72 characters wide, but for TV display with a single monitor in front of the classroom, a 40 character line is about the limit of resolution for the students in the back.

The programming language must be one which allows interactive execution of the program. In other words, one wishes to be able to call the program into execution, give it some input, obtain results as soon as they are calculated, then give it more input and so forth, all in a single run of the program. BASIC is such a language and is available on nearly all time-sharing systems. Other time-sharing languages which can also be used in this manner include FOCAL (PDP), TFOR (RCA), and Dial 360 FORTRAN (IBM360). These languages share both advantages and disadvantages for demonstration program use. Firstly, one writes interactively in these languages, so program modification and debugging are relatively easy and fast. In the second place, input is relatively free of format requirements, thus simplifying the running of programs. However, these languages are limited in computing power and program size compared to batch-oriented languages such as FORTRAN IV. Moreover, they tend to be relatively slow in execution. These limitations are perhaps most acute for quantum chemistry demonstration programs. Among the programs discussed below, DIATH2 ($H_2$ binding energy) uses double precision arithmetic in three of its subprograms, and SHMO (Huckel calculations) diagonalizes matrices up to 25 x 25. Since double precision is not available in BASIC, and since a 25 x 25 matrix diagonalization would consume excessive CPU time in BASIC or TFOR, it was necessary for the author to use standard FORTRAN IV for both programs, given the limitations of the RCA Spectra 70/46 then in use at WPI. The use of FORTRAN IV, however, has its own disadvantages. Some computers (notably IBM 360) do not allow interactive execution of FORTRAN IV. This limitation does not apply to PDP, CDC, UNIVAC, or RCA time-sharing systems, however. The writing and testing of programs is more cumbersome than with a language like BASIC. Errors in input can often cause complete termination of execution. Moreover, input is subject to fairly rigid format specifications.

The author has recently converted a number of his programs for operation on a DEC System 1050 computer. Since this machine was designed primarily as a central processor for a time-sharing network, its standard FORTRAN IV has some very useful terminal-oriented features. Program files may be constructed either by reading in cards or by typing statements into a remote terminal. Several powerful file editors are available for making corrections or modifications to programs, and for facilitating conversion of programs from other systems. Although line-by-line interpretation is not available, compilation is rapid and can be initiated from a terminal with immediate output of diagnostic messages (or execution results). Finally, free format input is available.

The choice of a computer language for a given demonstration program depends largely on the above stated factors. An instructor developing a program from scratch would probably write it in BASIC or another of the interactive programming languages. He would subsequently convert to FORTRAN IV only if he found he needed extra computation power unavailable in the interactive language. On the other hand, if he had an already working batch FORTRAN IV program, he would probably do little more than modify its input and output for effective interactive execution.

Regardless of language, the effectiveness of a demonstration program depends rather critically on how well its input and output are designed. As suggested above, output should be held to a minimum. At the same time, however, it should be made as clear and legible as possible. Numerical results should be labeled or grouped into columns with headings. Typewriter graphics are often quite effective. The author has found it useful to write his programs so they can output their own input instructions at the discretion of the user, as well as a title which tells what the program does. Often programs will be written that can deliver a number of different kinds of results from a single set of input parameters. In such cases it is very useful for the instructor to be able to tell the computer not to type the results he whishes not to discuss. As for input, a seemingly trivial, but nonetheless important point is to avoid inputting the same value of a given parameter more than once. For example, a program for calculating pressures according to the van der Waals law should be capable of outputting as many different pressures of the same gas for different V, n. or T without requiring repeated input of the parameters a and b.

In the event the programming language requires strictly formatted input it is extremely useful to have the computer type a line of characters just above the input line which will serve to indicate the specified input format. For example, in program TITRATE (described below) the line "GGGGGGGGG GGGGGGGGG I" is typed to indicate that the acid and base concentrations and number of dissociation constants are to be entered in FORMAT (2G10.7, I2). It should be noted that integer input (I format) and the exponent part of E format must be right justified, while floating point numbers (F format) need only fall somewhere within the input field.

Finally, one should attempt to program in such a way as to make one's programs transportable. The author, when writing programs to run on an RCA Spectra 70/46 or a DEC system 1050, attempts to make the conversion job to IBM, CDC, UNIVAC, etc., equipment as trivial as possible. One aspect of this is to avoid nonstandard use of programming languages which one's own machine might support but others do not. For example, although RCA BASIC accepts the statement A = B, this author uses the form: LET A = B, which is universally acceptable to BASIC interpreters. Another good practice is to use integer variables to identify input and output devices in FORTRAN READ and WRITE statements.

For the instructor who is contemplating the use of demonstration programs, transportability can be quite important. Many programs have been developed, and most of them are available upon request to their authors or in some cases to program exchanges. The Eastern Michigan University Center for the Exchange of Chemistry Computer Programs (Ronald W. Collins, Department of Chemistry, Eastern Michigan University, Ypsilanti, Michigan) distributes documentation of a large number of chemical education programs. The Quantum Chemistry Program Exchange, Indiana University (Chemistry Department, Room 204, Bloomington, Indiana, 47401), distributes documentation and machine-readable copies of this author's programs DIATH2 and TITRATE. This availability of tested programs makes it possible for an instructor to use demonstration programs in his classroom without having to write his own programs from scratch. Refer to Table 1 for a summary of the demonstration programs discussed in this chapter.

TABLE 1

Demonstration Programs Discussed in This Chapter

| Program | Calculates | Language | Authors | Availability |
|---|---|---|---|---|
| DIATH2 | Binding energies of $H_2$ | FORTRAN IV | P. E. Stevenson | QCPE[a] |
| TITRATE | Acid-base titration curves | FORTRAN IV | P. E. Stevenson, J. E. Merrill & B. R. Thompson | QCPE[a] |
| BOLPLOT | Statistical distributions | FORTRAN IV | C. R. Williams & P. E. Stevenson | QCPE[a] |
| CH357A3 | Equilibrium constants | TFOR | P. E. Stevenson | Not available[b] |
| VDW2 | Vapor pressures | BASIC | D. R. Lyons & P. E. Stevenson | From author[c] |
| SHMO | Huckel calculations | FORTRAN IV | P. E. Stevenson | Not available |
| NEMO | NEMO calculations | FORTRAN IV | F. P. Boer & P. E. Stevenson | Not available |
| GASLAW | Van der Waals law pressures | FORTRAN IV | J. E. Merrill | Not available[b] |
| FEF | Free energies vs T | FORTRAN IV | J. E. Merrill | Not available[b] |

[a] For information, write to Quantum Chemistry Program Exchange, Room 204, Chemistry Department, Indiana University, Bloomington, Indiana, 47401.

[b] Program will be made available to QCPE when ready.

[c] Write to P. E. Stevenson, Department of Chemistry, Worcester Polytechnic Institute, Worcester, Massachusetts 01609. Source code will be sent on paper punch tape generated on a DEC system 1050.

## III. EDUCATIONAL USE OF DEMONSTRATION PROGRAMS

The primary use of demonstration programs is to produce numerical or graphical results of complicated computations beyond the reach of mental arithmetic or slide rule calculations for immediate display to a class. The use of such results in class discussions largely determines the value of the demonstration program. It is temptingly easy to let a demonstration calculation replace a good explanation of a physical law which can be expressed mathematically--but the demonstration program should rather free the instructor to undertake a deeper and more thorough explanation and exploation of the law and its implications. Availability of demonstration programs should also enable the instructor to make use of somewhat more realistic and complex models in his descriptions of phenomena to his classes. Another kind of demonstration program is useful in discussions of ionic equilibria. The equations describing most ionic equilibrium situations are nearly intractable without the aid of approximations based on chemical intuition. However, it is most difficult to teach students how to make the "right" approximations. The author's program, TITRATE, can compute concentrations of all substances involved in an acid-base equilibrium when given amounts of acid and base and values of all dissociation constants. From those results it is easy for the instructor to show his class which quantities can be neglected in the original equations and thus how to do the calculation by hand.

In addition to classroom use, demonstration programs can be used outside of class. If there is direct student access to the college time-sharing system, students may be given access to these programs for optional extra study or even homework assignments. More importantly, students with an inclination toward programming can be easily enlisted to write demonstration programs. The students who write such programs usually learn enough from the exercise that their time is well spent even if their programs are never used.

With these demonstration programs (unlike CAI tutorial programs) it is almost impossible to specify connect times. Fifty minutes is typical for a program being used in a classroom demonstration, given that the instructor wishes to conduct a thorough discussion of the computed results. However, for all of the programs described below, computation of a single case need take no more than one or two minutes of connect time. Hence the time a student spends at a teletype running one of these programs depends almost entirely on the number of different cases he tries, the amount of time he "pauses" while the computer is ready to accept input, and even to some extent on his typing speed.

## IV. DEMONSTRATION PROGRAMS WRITTEN BY THE AUTHOR AND HIS STUDENTS

### A. DIATH2--A Program for Demonstrating Chemical Bonding in Hydrogen

Language: FORTRAN IV (RCA)
Author: P. E. Stevenson
Available from QCPE (Program #182)

$H_2$ and $H_2^+$ are traditionally the first molecules encountered by the student of quantum chemistry. At some point in his introduction to the subject, he sees his instructor dutifully derive formulas (which can be found in any standard introductory quantum chemistry text, e.g., L. Pauling and E. B. Wilson, Jr., Introduction to Quantum Mechanics, McGraw-Hill, New York, 1935, pp. 326-353.) for the energies of the valence bond function for $H_2$ [1] and the molecular orbital functions of $H_2^+$ [2] and $H_2$ [3]. He may even encounter the Weinbaum [4] and configuration interaction treatments. However, when it comes to numerical results, all he encounters are tables and perhaps a dissociation curve or two. He never sees nor participates in the process of calculating these results. If for this lack of participation he concludes that quantum mechanics is a "black art," he is not without some justification.

However, with the aid of a computer and the program DIATH2 described here (Fig. 1), it is possible for the quamtum chemistry teacher to add numerical results to his classroom discussion of $H_2$ and $H_2^+$. Using a remote terminal connected to a time-sharing computer, he may enter arbitrary values of the internuclear distance R and 1s orbital exponent $\zeta$ and obtain the resulting energies within seconds. He could enter the "book" values of R and $\zeta$, or with somewhat more expenditure of time, search for their optimum values. He could assign the construction of optimized dissociation curves as homework.

The program treats the various wave functions for $H_2$ and $H_2^+$ which can be constructed from 1s orbitals centered on each nucleus, namely, the valence bond function of $H_2$, the molecular orbital functions for $H_2^+$ and $H_2$, and the simplest configuration interaction function for $H_2$. The Weinbaum treatment is omitted, since it is formally identical to the configuration interaction treatment. The program, written entirely in FORTRAN IV, takes the input values of R and $\zeta$, converting R into atomic units, if necessary, and evaluates all the necessary integrals involving components of the Hamiltonian operators over 1s orbitals [Eq. (1)] on the two nuclei:

$$1s = \zeta^{3/2}\pi^{-1/2}e^{-\zeta\rho} \qquad (1)$$

The program then evaluates electronic, total, and binding energies. The valence bond wave function, the molecular orbital wave functions for $H_2^+$ and $H_2$, and the configuration interaction wave function are all treated by the program (the specific formulas are not repeated here since they are readily available). At the user's option any or all of the output may be typed at the terminal for immediate inspection. Any output not typed is printed off-line for later use. An auxiliary subroutine "PARAB," is available which constructs a parabolic curve when given three points. It also gives the extreme point of the curve. This is useful when one is optimizing for $\zeta$ or R and desires a quadratic interpolation. The use of the input distance R = 0.0 results in output of the electronic energies of $He^+$ and He.

The author has used this program in his "Molecular Orbital Theory" course. Student enthusiasm was high, and a number of points were brought out and clarified which are often confusing or not appreciated. These points included the incorrect dissociation of the molecular orbital function, the united atom picture for electronic energy, and the role of orbital contraction. Application of the variation theorem was rather effectively demonstrated. However, in the course of an 80-minute class period, there was only time enough for a thorough discussion of $H_2^+$. Thus, in future years he plans to assign homework problems involving student use of the program, as well as using it for demonstration purposes.

## B. TITRATE--Ionic Equilibria

Language: FORTRAN IV (RCA and PDP)
Authors: P. E. Stevenson, J. E. Merrill, and B. R. Thompson
Available from QCPE (Program #196)

Ionic equilibrium is traditionally one of the most difficult topics in the freshman chemistry syllabus. The combination of complex relationships among the variables, the need for such complicated mathematical techniques as the method of successive approximations, and the need to recognize which variables can be neglected in which equations proves to be the downfall of many a student. Demonstration calculations, though pedagogically necessary, are tedious and time-consuming for an instructor to perform in class.

The author has written a demonstration program which simulates acid-base titrations. This program, named TITRATE, computes a pH vs volume of base added curve for the titration of a monobasic, dibasic, or tribasic weak acid with a strong base, or the analogous curve for the titration of a monoacidic, diacidic, or triacidic weak base with strong acid. Program TITRATE can also give concentrations of all substances at any point in the titration.

```
 EX DIATH2.F4
FORTRAN:  DIATH2.F4
LOADING

DIATH2 5K CORE
EXECUTION


HYDROGEN MOLECULE AND MOLECULE ION ENERGIES (IN A. U.)
DATA CONSIST OF 5 CONTROL FLAGS PLUS THE ORBITAL EXPONENT
AND INTERNUCLEAR DISTANCE.
THE FLAGS ARE:
 UNITS (U = 0 FOR BOHRS, U = 1 FOR ANGSTROMS,
        U = 2 TO CALL PARAB, U = 99 TO STOP)
 INTEGRAL OUTPUT (I = 1 TO TYPE)
 VALENCE BOND OUTPUT (V = 1 TO TYPE)
 MOLECULAR ORBITAL OUTPUT (M = 1 TO TYPE)
 CONFIGURATION INTERACTION OUTPUT (C = 1 TO TYPE).
INPUT FORMAT IS I2,4I1,2F
 UIVMC ZZZZZZZZZ RRRRRRRRR   RUN  1
 11111 1.0       0.740

HYDROGEN INTEGRALS AT   1.39842 BOHRS  WITH ZETA = 1.0000
SAB   =   0.7533800530
TAA   =   0.5000000000
TAB   =   0.2156895230
VAARA =   1.0000000000
VABRA =   0.5923795480
VAARB =   0.6104687300
EAAAA =   0.6250000000
EAABB =   0.5037293730
EAAAB =   0.4262085260
EABAB =   0.3237187120

  ENERGIES    ELECTRONIC  TOTAL   BINDING(EV)
VALENCE BOND     -1.8204   -1.1053    2.86570
MO: H2+          -1.1860   -0.4709   -0.79121
    SIGMAG**2    -1.8059   -1.0908    2.47071
    SIGMAU**2    -0.5603    0.1548
    CI            0.1402
    PSI-1        -1.8215   -1.1064    2.89492
    PSI-2        -0.5447    0.1704
```

Fig. 1. Sample input to and output from DIATH2. Run 1 has for input the experimental bond distance and the orbital exponent 1.0. All possible output is typed. Run 2 inputs the bond distance and orbital exponent which optimize the configuration interaction problem. Only the output pertinent to CI is typed. Runs 3 and 4 illustrate the use of DIATH2 for calculations on $He^+$ and He. (Note: this figure and all the others were generated from DEC System 10 versions of the author's programs. All input data are underlined.)

```
PSI-1 =  .994(SIGMAG**2)  -  .111(SIGMAU**2)
PSI-2 =  .111(SIGMAG**2)  +  .994(SIGMAU**2)
 UIVMC ZZZZZZZZZ RRRRRRRRR    RUN  2
 10001 1.193       0.77


  ENERGIES     ELECTRONIC   TOTAL   BINDING(EV)
MO: H2+            -1.2316   -0.5444     1.20809
     SIGMAG**2    -1.8143   -1.1271     3.45720
     SIGMAU**2    -0.3656    0.3216
     CI            0.1747
     PSI-1        -1.8351   -1.1478     4.02211
     PSI-2        -0.3448    0.3424


PSI-1 =  .993(SIGMAG**2)  -  .118(SIGMAU**2)
PSI-2 =  .118(SIGMAG**2)  +  .993(SIGMAU**2)
 UIVMC ZZZZZZZZZ RRRRRRRRR    RUN  3
 00010 2.0          0.0


  ENERGIES     ELECTRONIC   TOTAL   BINDING(EV)
   HE+              -2.0000
   HE               -2.7500
 UIVMC ZZZZZZZZZ RRRRRRRRR    RUN  4
 00010 1.6875      0.0


  ENERGIES     ELECTRONIC   TOTAL   BINDING(EV)
   HE+              -1.9512
   HE               -2.8477
 UIVMC ZZZZZZZZZ RRRRRRRRR    RUN  5
 99


CPU TIME: 1.14   ELAPSED TIME: 4:27.75
NO EXECUTION ERRORS DETECTED

EXIT
```

Fig. 1 - Continued

One might better appreciate the value of a simulated titration if one recalls that titrations unify the problems of weak acid and base pHs, buffers, and hydrolysis. A detailed discussion of the titration of a monobasic weak acid with strong base will help clarify this. One deals with the equilibrium of the weak acid HA with its ions (this discussion is limited to the case of the monobasic weak acid HA; however, the program algorithm is capable of handling $H_2A$ and $H_3A$ as well):

$$HA(aq) = H^+ + A^- ; \tag{2}$$

$$\frac{[H^+]\,[A^-]}{[HA]} = K_a ; \tag{3}$$

the self-dissociation of water:

$$H_2O = H^+ + OH^- , \tag{4}$$

$$[H^+]\,[OH^-] = K_w = 1.0 \times 10^{-14} ; \tag{5}$$

the conservation of the initial amount of acid:

$$[HA] + [A^-] = F_a\left(\frac{V_a}{V_a + V_b}\right) ; \tag{6}$$

and the conservation of electric charge:

$$[H^+] + [Na^+] = [OH^-] + [A^-]. \tag{7}$$

The $Na^+$ ion concentration is known to be

$$[Na^+] = F_b\left(\frac{V_b}{V_a + V_b}\right) . \tag{8}$$

Equations (6), (7), and (8) assume that a given point in the titration is attained by mixing $V_a$ ml of weak acid of formality $F_a$ with $V_b$ ml of NaOH of formality $F_b$. The program, in fact, assumes $V_a = 50.0$ ml and accepts $F_a$, $F_b$, and $K_a$ as input parameters. $V_b$ is the independent variable in the titration and is automatically incremented from 0.0 ml to a point 20% past the equivalence point. The algebraic solution of Eqs. (3), (5), (6), (7), and (8) is programmed. The variables $[OH^-]$ and $[A^-]$ are thus eliminated from Eq. (7) and the resulting cubic equation in $(H^+)$ is solved by the Newton-Raphson method.

One would, of course, never go through this complete solution in a hand calculation, but would instead neglect certain quantities in Eqs. (6) and (7) and then plug the equations into Eq. (3), using the method of successive approximations as a last resort. However, most freshman lack the chemical intuition to be able to know what quantities to neglect and when to neglect them. Since program TITRATE calculates the concentrations of all the ions in solution without neglecting any of them, a freshman chemistry instructor can use it in class to give his students some feeling for the numbers resulting from ionic equilibrium calculations. Thus, students could begin to acquire the chemical intuition which they need in order to master the subject of ionic equilibrium.

The author has used program TITRATE experimentally in his two recitation sections in CH102 (freshman chemistry) at WPI. An ASR-33 teletype was set up at the side of a 30-seat classroom using a TV camera and monitor to show the results to the class. Communication with the RCA Spectra 70/46 computer at the Worcester Area College Computation Center was by telephone. He spent approximately ten minutes at the beginning of class discussing Eqs. (2) through (8). Then he obtained the titration curve of 50.0 ml of 0.1 F acetic acid with 0.1 F NaOH (Fig. 2) and drew attention to the equivalence point pH of 8.72. The specific problems of the weak acid solution, the 1:1 buffer, and the hydrolysis of the salt of a weak acid problem were then demonstrated by requesting computation of the concentrations of all species in solution at 0.0 ml, 25.0 ml, and 50.0 ml, respectively, in the titration. For these problems, all the standard approximations are valid, and the program convinced the students that the concentrations of neglected species are indeed at least several orders of magnitude smaller than those of the major species. There remained time for only one additional titration. In one class, titration of a moderately strong weak acid ($K_a = 5 \times 10^{-3}$) was demonstrated; in the other the titration of a very very weak acid ($K_a = 1.0 \times 10^{-12}$) was demonstrated. In these two cases at least one approximation breaks down in each of the standard problems. For example, in the hydrolysis of the salt of the very very weak acid, the self-ionization of water [Eqs. (3) and (4)] is significant. These breakdowns were clearly demonstrated.

Subsequently the program has been used as a student-run supplement to a sophomore laboratory exercise in which an unknown weak acid or base is titrated while being monitored with a pH meter. From the resulting pH vs volume curve the student is to calculate the molecular weight and pK values. He can then use these values if he wishes as program input to check his work.

## C. BOLPLOT--A Simple Demonstration of the Statistical Distribution of Energies among Identical Particles

Language: FORTRAN IV (PDP)
Authors: P. E. Stevenson and C. R. Williams
Will be submitted to QCPE

Statistical mechanics forms a very important conceptual bridge between the quantum properties of atoms and molecules and the thermodynamic properties of matter in bulk. The basic concepts of statistical mechanics should therefore be taught as part of the freshman course in chemistry.

The basic principles are simple enough. A gas can be modeled as a collection of indistinguishable microscopic particles, each one having a known spectrum of quantum states with characteristic energies. Quantum

```
R TITRE


THIS PROGRAM SIMULATES THE TITRATION OF A WEAK ACID WITH A STRONG BASE
OR A WEAK BASE WITH A STRONG ACID.
ENTER "ACID", "BASE", OR "STOP".
ABAB
ACID

TITRATION OF 50.0 ML SAMPLE OF WEAK ACID WITH STRONG BASE
ENTER ACID CONCENTRATION, BASE CONCENTRATION,
AND NUMBER OF DISSOCATION CONSTANTS: (FREE FORMAT, SEPARATE WITH COMMAS)
0.1,0.1,1

ENTER DISSOCIATION CONSTANTS: (FREE FORMAT, ONE PER LINE)
1.75E-05

INDICATE NUMBER OF POINTS ON TITRATION CURVE (MANY/FEW/NONE/PLOT)
ABAB
PLOT

 50.000   TO FIRST ENDPOINT
 VOLUME   PH  2    3    4    5    6    7    8    9   10   11   12
  0.000   2.88     *
  5.000   3.81          *
 10.000   4.16            *
 15.000   4.39             *
 20.000   4.58              *
 25.000   4.76               *
 30.000   4.93                *
 35.000   5.13                 *
 40.000   5.36                  *
 45.000   5.71                    *
 50.000   8.73                                   *
 55.000  11.68                                                 *
 60.000  11.96                                                   *
FOR CONCENTRATIONS OF ALL SPECIES AT ANY VOLUME IN THE TITRATION,
ENTER 1 AND THE VOLUME.  OTHERWISE ENTER 0
I **VOL**
1 0.0
```

```
  (H+)       (OH-)      (NA+)
1.314E-03 7.609E-12 0.000E-01

  (HA)       (A-)
9.869E-02 1.314E-03

I **VOL**
1 25.0

  (H+)       (OH-)      (NA+)
1.748E-05 5.720E-10 3.333E-02

  (HA)       (A-)
3.332E-02 3.335E-02

I **VOL**
1 50.0

  (H+)       (OH-)      (NA+)
1.871E-09 5.346E-06 5.000E-02

  (HA)       (A-)
5.344E-06 4.999E-02

I **VOL**
0

ENTER "ACID", "BASE", OR "STOP".
ABAB
STOP


CPU TIME: 1.11  ELAPSED TIME: 3:59.17
NO EXECUTION ERRORS DETECTED

EXIT
```

Fig. 2. Sample input to and output from TITRATE. In this example 0.1 F acetic acid is titrated with 0.1 F NaOH. A plot of the titration curve is generated. (The options MANY and FEW give more points, especially near equivalence points.) Then concentration data are generated for pure acetic acid (0.0 ml), the 1:1 buffer (25.0 ml), and hydrolysis of 0.05 F sodium acetate (50.0 ml). (Note that the DEC version of this program is named "TITRE.")

states for the gas as a whole can in principle be specified by giving the numbers of particles in each particle state. However, the number of such states for the system as a whole is, in practical terms, uncountably large.

At this point statistical mechanics comes to the rescue by postulating that only the average distribution of particles need be considered for purpose of calculating bulk properties. In advanced courses this can be proven to be the Boltzmann distribution;

$$N_i/N = P_i = e^{-\epsilon_i/kT}/Q \ , \qquad (9)$$

where $N_i$, $P_i$, and $\epsilon_i$ are, respectively, the number of particles in the $i^{th}$ particle state, the probability of a given particle being in the $i^{th}$ state, and the energy of the $i^{th}$ state. N is the total number of particles, k is the Boltzmann constant, T is the absolute temperature, and Q is a constant which makes all the probabilities add to 1. Note that if there are several states with energy $\epsilon_i$, Eq. (9) refers to them one at a time.

The derivation of Eq. (9) is beyond the understanding of most freshmen, yet an appreciation of the implications of the Boltzmann distribution is rather important for their understanding of chemistry. The author feels that it is important to discuss it, but unsatisfactory merely to state it without further justification. The problem of justifying the Boltzmann distribution has been solved by Plumb [5]. He considers a system that consists of a very small number (three, for example) of gas particles which possesses a fixed small number of units of energy (six, for example). He assumes that the single particle states are quantized at $\epsilon = 0, 1, 2, 3...$ units of energy. Then he generates explicitly all the allowed distributions and calculates the average numbers of particles in each particle state [corresponding to $N_i$ in Eq. (9)]. He also calculates the probability (corresponding to $P_i$). A sample calculation is shown in Table 2 for three particles and six units of energy. He generates several distributions explicitly, then passes in principle to the limiting behavior of the distribution as the system grows to large numbers of particles while the average energy per particle remains fixed. Even for as small a system as that illustrated in Table 2, the distribution bears some resemblance to an exponential function, and students readily accept that the limiting behavior is exponential.

Hand calculations with Plumb's model are too time-consuming for classroom use for systems of more than three particles and six units of energy. However, program BOLPLOT can be used to extend the explicit calculations to systems of five particles and twenty-five units of energy. As can be seen from the sample output in Fig. 3, a system of as few as five particles has a clearly exponential-like distribution.

```
R MAXBOL


PROGRAM TO CALCULATE THE DISTRIBUTION FUNCTION OF A
MICROCANONICAL ENSEMBLE OF INDISTINGUISHABLE PARTICLES
ENTER (AS INTEGERS) THE NUMBER OF PARTICLES (UP TO 5)
AND TOTAL ENERGY (UP TO 25 UNITS)
 NNNN EEEE
    5   10

ONE DIMENSIONAL DISTRIBUTION OF 5 PARTICLES OF AVERAGE ENERGY  2.00
E(I) N(I)    P(I) .0        .1        .2        .3        .4        .5
 0   1.467   0.293 :*********:*********:*********
 1   1.100   0.220 :*********:*********:**
 2   0.900   0.180 :*********:********
 3   0.533   0.107 :*********:*
 4   0.367   0.073 :*******
 5   0.233   0.047 :*****
 6   0.167   0.033 :***
 7   0.100   0.020 :**
 8   0.067   0.013 :*
 9   0.033   0.007 :*
10   0.033   0.007 :*

THE NUMBER OF POSSIBLE DISTRIBUTIONS (ND) IS   150    LN(ND) =   5.01
DO YOU WISH TO SEE THE THREE DIMENSIONAL DISTRIBUTION? (Y/N) Y


MAXWELL-BOLTZMANN (THREE DIMENSIONAL) DISTRIBUTION OF 5 PARTICLES
OF AVERAGE ENERGY  6.00 (DIRECT COMBINATION OF PROBABILITIES)
E(I) P(I) .0        .1        .2        .3        .4        .5
 0   0.025 :***
 1   0.057 :******
 2   0.089 :*********
 3   0.108 :*********:*
 4   0.115 :*********:*
 5   0.111 :*********:*
 6   0.102 :*********:
 7   0.089 :*********
 8   0.075 :*******
 9   0.061 :******
10   0.048 :*****
11   0.037 :****
12   0.027 :***
13   0.019 :**
14   0.013 :*
15   0.009 :*
16   0.006 :*
 NNNN EEEE
    5   25
```

Fig. 3. Sample input to and output from BOLPLOT. Systems of five particles are given, respectively, 10 and 25 units of energy. The exponential nature of the distribution and the influence of temperature are both illustrated. Since writing the text of this chapter, the author has substantially revised this program. It now gives, in addition to the one-dimensional Boltzmann-like distribution, a three-dimensional Maxwell-Boltzmann-like distribution. The three-dimensional probability $P_i$ that a particle will have energy $E_i$ is given by:

```
ONE DIMENSIONAL DISTRIBUTION OF 5 PARTICLES OF AVERAGE ENERGY  5.00
E(I) N(I)    P(I) .0         .1         .2         .3         .4         .5
 0  0.700  0.140 :*********:****
 1  0.631  0.126 :*********:***
 2  0.554  0.111 :*********:*
 3  0.493  0.099 :*********:
 4  0.427  0.085 :*********
 5  0.379  0.076 :********
 6  0.318  0.064 :******
 7  0.273  0.055 :*****
 8  0.225  0.045 :*****
 9  0.191  0.038 :****
10  0.156  0.031 :***
11  0.133  0.027 :***
12  0.106  0.021 :**
13  0.090  0.018 :**
14  0.072  0.014 :*
15  0.061  0.012 :*
16  0.048  0.010 :*
17  0.040  0.008 :*
18  0.029  0.006 :*
19  0.024  0.005 :
20  0.016  0.003 :
21  0.013  0.003 :
22  0.008  0.002 :
23  0.005  0.001 :
24  0.003  0.001 :
25  0.003  0.001 :

THE NUMBER OF POSSIBLE DISTRIBUTIONS (ND) IS  1885    LN(ND) =   7.54
DO YOU WISH TO SEE THE THREE DIMENSIONAL DISTRIBUTION? (Y/N) Y
```

Fig. 3 - Continued. $P_i = \sum_{j+k+l=i} P_j P_k P_l$, where $P_j$, $P_k$, and $P_l$ are the one-dimensional probabilities already calculated, and the summation is over all values of j, k, and l that add up to i. The average energy of a particle in three dimensions is necessarily three times the one-dimensional average. The logarithm of the number of one-dimensional distributions is also calculated ln (ND) and may be related to the entropy of the essemble. Because of these changes, the program has been renamed "MAXBOL."

Program BOLPLOT has Plumb's algorithm programmed explicitly into it. (At no point does the program call upon EXP.) Its input consists of the number of particles and the total energy of the system. Its output consists of tabulated values of probabilities vs energy. The probabilities are also plotted.

TABLE 2

Plumb's Model of the Distribution of Six Units of Energy among Three Particles. Entries in the Table Are Numbers of Particles

| Distributions | Particle energy states | | | | | | |
|---|---|---|---|---|---|---|---|
| | 0 | 1 | 2 | 3 | 4 | 5 | 6 |
| System state 1 | 2 | 0 | 0 | 0 | 0 | 0 | 1 |
| System state 2 | 1 | 1 | 0 | 0 | 0 | 1 | 0 |
| System state 3 | 1 | 0 | 1 | 0 | 1 | 0 | 0 |
| System state 4 | 0 | 2 | 0 | 0 | 1 | 0 | 0 |
| System state 5 | 1 | 0 | 0 | 2 | 0 | 0 | 0 |
| System state 6 | 0 | 1 | 1 | 1 | 0 | 0 | 0 |
| System state 7 | 0 | 0 | 3 | 0 | 0 | 0 | 0 |
| Average | 5/7 | 4/7 | 5/7 | 3/7 | 2/7 | 1/7 | 1/7 |
| Probability | 0.24 | 0.19 | 0.24 | 0.14 | 0.10 | 0.05 | 0.05 |

Classroom discussions of the Boltzmann distribution should begin with one or two hand calculations, such as in Table 2, in order that the students understand the principles of the model. Then BOLPLOT can be used to make the passage to the limit of large systems much more readily acceptable to the students.

## D. CH357A3--Statistical Mechanical Calculation of Chemical Equilibrium Constants

Language: FORTRAN IV (PDP)
Author: P. E. Stevenson
Will be submitted to QCPE

It is possible to use the principles of statistical mechanics to derive equilibrium constants of chemical reactions. If one assumes that the reagents A, B, and C are ideal gases and form an ideal mixture, and that they undergo the reaction

$$A + B \rightleftarrows C , \tag{10}$$

then the equilibrium constant for the reaction is

$$K_N = \frac{N_C}{N_A N_B} = \frac{Q_C}{Q_A Q_B} e^{-\Delta\epsilon /kT} . \tag{11}$$

```
MAXWELL-BOLTZMANN (THREE DIMENSIONAL) DISTRIBUTION OF 5 PARTICLES
OF AVERAGE ENERGY 15.00 (DIRECT COMBINATION OF PROBABILITIES)
E(I) P(I) .0          .1         .2         .3         .4         .5
 0   0.003 :
 1   0.007 :*
 2   0.013 :*
 3   0.020 :**
 4   0.026 :***
 5   0.032 :***
 6   0.038 :****
 7   0.042 :****
 8   0.046 :*****
 9   0.049 :*****
10   0.050 :*****
11   0.051 :*****
12   0.051 :*****
13   0.050 :*****
14   0.049 :*****
15   0.047 :*****
16   0.045 :****
17   0.042 :****
18   0.039 :****
19   0.036 :****
20   0.033 :***
21   0.030 :***
22   0.027 :***
23   0.025 :**
24   0.022 :**
25   0.019 :**
26   0.017 :**
27   0.015 :*
28   0.013 :*
29   0.011 :*
30   0.009 :*
31   0.008 :*
32   0.007 :*
33   0.005 :*
 NNNN EEEE
     0

CPU TIME: 24.90 ELAPSED TIME: 7:29.07
NO EXECUTION ERRORS DETECTED

EXIT
```

Fig. 3 - Continued

$K_N$ is the equilibrium constant in terms of numbers of molecules; $N_A$, $N_B$, and $N_C$ are the equilibrium numbers of molecules of species A, B, and C; $Q_A$, $Q_B$, and $Q_C$ are the molecular partition functions; $\Delta\epsilon$ is the difference between the atomization energy of C and those of A and B; k is the Boltzmann constant; and T is the absolute temperature. The ideal gas law can be used to derive the equilibrium constant $K_p$ in terms of partial pressures:

$$K_p = \frac{(RT/NV)q_C}{(RT/NV)q_A\,(RT/NV)q_B}\; e^{-\Delta\epsilon/kT}, \tag{12}$$

where R is the gas constant, N is Avogadro's number, and V is the volume of the system. Equation (2) is independent of volume now, since each q is first-order in V. The partition functions $q_A$, $q_B$, and $q_C$ are in turn functions of atomic or molecular weights (translational motion), moments of inertia and symmetry number (rotational motion), normal vibration frequencies, and the degeneracies and energies of the ground and low-lying excited electronic states.

Equation (12) is fairly simple in principle, but can be enormously difficult to use for calculations. The combined problem of keeping units, signs, and zeroes of energy consistent usually results in computational errors. Program CH357A3 makes it possible to compute equilibrium constants by Eq. (12) with all the consistency problems worked out within the program. It is set up to handle reactions among atomic and diatomic ideal gases with up to five different species present in the reaction. ($H_2$, HD, and $D_2$ are excluded, since the rotational partition function is assumed to be classical.) The input data (Figs. 4, 5) consist of the species name (4 characters), the stoichiometric coefficient NSTO (negative for reactants, positive for products), masses of the two atoms, M1 and M2, (M2 = 0 tells the program that the species is an atom), degeneracies of the ground and lowest excited electronic states, G1 and G2, energy of the lowest excited electronic state, EELEC2 (set G2 = 0 and EELEC2 = 0 in order to disregard excited states), and if the species is a molecule the internuclear distance R, the vibration frequency NU, and the molecular dissociation energy DO, as measured from the lowest vibrational state. Input of NAME=END causes transfer of control to the next section of the program. The program then asks for a series of temperatures and pressures and for each pair computes E, $C_V$, S, and q for each species and $\Delta H$, $\Delta S$, $\Delta G$, and $K_p$ for the reaction. KEQ is $K_p$ calculated by Eq. (12); KG is $K_p$ calculated from $\Delta G$. To terminate execution, one enters a negative value of T.

## E. VDW2--Vapor Pressures from van der Waals Isotherms

Language: BASIC (PDP)
Authors: D. R. Lyons and P. E. Stevenson
Available directly from author on punched paper tape.

Classroom discussions of equations of state for nonideal gases are generally inhibited by computational difficulties. Typically, the van der Waals law

$$\left(P + \frac{n^2 a}{V^2}\right)(V - nb) = nRT \tag{13}$$

is written on the blackboard, and a is said to account for the attractive forces between molecules while b represents their finite size. Perhaps a homework problem or two might be assigned, but little else is normally done.

Thus, the students are denied exposure to some of the more interesting consequences of the van der Waals law. One such consequence is its prediction that there will be two phases (vapor and condensed) at temperatures below the critical temperature. For example, the van der Waals isotherm for 1 mole of $N_2$ at 100° K (P vs V curve) is shown in Fig. 6. The behavior of curve ABCDE is of special interest. Note that from B to D the pressure falls in response to a decrease in volume, a physically impossible result. Through thermodynamic arguments which will not be discussed here [6] it can be shown that the actual physical behavior of the system corresponds to the straight line ACE, along which the pressure is constant. So if one starts with $N_2$ gas at F on the curve and compresses it, the pressure will rise until point E is reached. Then a great amount of additional compression can occur with no increase in pressure until point A is attained. Further compression takes place only upon application of enormous pressures. What happens is that in going from E to A, the $N_2$ gas condenses to the liquid phase. The condensation pressure is determined by the requirement that the areas ABC and CDE must be equal. This condensation pressure is of course also the vapor pressure of the liquid at the particular temperature for which the isotherm was calculated.

Program VDW2 calculates vapor pressures from van der Waals isotherms. In order to use the program, one inputs values of a (in $l.^2$-atm/$mole^2$) and b (in l./mole), and then a series of temperatures (in °K). The program determines whether or not T is below the critical temperature, and if so, it calculates the phase transition pressure as outlined above and outputs it as the vapor pressure. If T is above the critical temperature, a message to that effect is typed. The algorithm is numerical, and for temperatures near the critical value it cannot sense the difference between the local minimum B and the local maximum D. In such cases the message "ERROR NO CHANGE" is typed. It is planned to substitute an algebraic algorithm to improve the performance of the program.

Program VDW2 (Fig. 7) demonstrates the possibility of calculating properties of real substances from the van der Waals equation of state. Unfortunately, the numerical results are not very good. For example, $N_2$ is predicted to boil at about 65° K, and water at 100° C has a vapor pressure of about 14 atm using the van der Waals law. The reason for the discrepancies is fairly clear, however. The van der Waals law is designed to be accurate in the region of the critical point, and the parameters a and b are computed from critical point data. This leads to serious errors in the calculated molar volume for the condensed phase, which is b according to the van der Waals law. For example, water in the liquid phase at room temperature has a molar volume of 0.018 l./mole, while b is about 0.03 l./mole. The result is that the van der Waals isotherm is in serious error in the region from B to A and beyond, the "well", ABC, being calculated narrower than it really

```
R EQUIB

STATISTICAL MECHANICAL CALCULATIONS OF EQUILIBRIUM CONSTANTS
OF REACTIONS INVOLVING ATOMS AND DIATOMIC MOLECULES.
UNITS ARE: ATOMIC MASS UNITS, ANGSTROMS, CM-1 (FOR EXCITATIONS)
ELECTRON VOLTS (FOR DISSOCIATIONS), AND KELVINS.

ENTER DATA FOR SPECIES 1

NAME=CL2

NSTO=-1

M1=35.453

M2=35.453

G1=1

G2=0

EELEC2=0

R=1.989

NU=564.9

D0=2.481

ENTER DATA FOR SPECIES 2

NAME=CL

NSTO=2

M1=35.453

M2=0

G1=4

G2=2

EELEC2=881
```

Fig. 4. Sample input to and output from CH357A3. The equilibrium constant for $Cl_2(g) \rightleftharpoons 2Cl(g)$ is calculated for T = 298°K. Species 1 is $Cl_2$ with a bond distance of 1.989 Å, vibration frequency of 564.9 $cm^{-1}$, and dissociation energy of 2.481 eV. Species 2 is Cl with a ground state of $^2P_{3/2}$ and lowest excited state (at 881 $cm^{-1}$) of $^2P_{1/2}$. The program was originally written in TFOR so it asked for input by typing the name of the variable on the terminal paper followed by an "equals" sign. It then waited for the user to type in the value of the input parameter. Now converted to DEC System 10 FORTRAN and renamed "EQUIB," its method of asking for input remains the same.

```
ENTER DATA FOR SPECIES 3

NAME=END


ENTER TEMPERATURE

T=298


RESULTS FOR CL2
E   =   1.593E+03 CAL/MOLE
CV  =   6.074E+00 CAL/MOLE/DEG
S   =   5.324E+01 CAL/MOLE/DEG
PQ  =      33.813

RESULTS FOR CL
E   =   9.059E+02 CAL/MOLE
CV  =   3.232E+00 CAL/MOLE/DEG
S   =   3.945E+01 CAL/MOLE/DEG
PQ  =      31.303

RESULTS FOR THE REACTION:
         1.0 CL2 =    2.0 CL

DELTAH =  5.804E+04 CAL
DELTAS =  2.566E+01 CAL/DEG
DELTAG =  5.040E+04 CAL
AT T   =     298.00 DEG

LOGKEQ = -36.965

KEQ    =  1.084E-37
```

Fig. 4 - Continued

is, so that the vapor pressure is predicted to be greater than the observed value. For "qualitative" demonstrations, the standard values of b will suffice, but for greater accuracy, one might wish to make b smaller, or even temperature-dependent.

## F. Other Demonstration Programs

### 1. Quantum Chemistry

Program SHMO performs simple Huckel molecular orbital calculations. Input consists of the matrix dimension, number of $\pi$ electrons, and the Hamiltonian matrix elements in units of $\beta$. Teletype output consists of

```
ENTER TEMPERATURE

T=1000

RESULTS FOR CL2
E  =  6.256E+03 CAL/MOLE
CV =  6.849E+00 CAL/MOLE/DEG
S  =  6.360E+01 CAL/MOLE/DEG
PQ =      35.879

RESULTS FOR CL
E  =  3.291E+03 CAL/MOLE
CV =  3.326E+00 CAL/MOLE/DEG
S  =  4.596E+01 CAL/MOLE/DEG
PQ =      32.672

RESULTS FOR THE REACTION:
         1.0 CL2 =    2.0 CL

DELTAH =   5.955E+04 CAL
DELTAS =   2.832E+01 CAL/DEG
DELTAG =   3.122E+04 CAL
AT T   =     1000.00 DEG

LOGKEQ =   -6.825

KEQ    =   1.497E -7

ENTER TEMPERATURE

T=2000

RESULTS FOR CL2
E  =  1.316E+04 CAL/MOLE
CV =  6.928E+00 CAL/MOLE/DEG
S  =  6.975E+01 CAL/MOLE/DEG
PQ =      37.154

RESULTS FOR CL
E  =  6.489E+03 CAL/MOLE
CV =  3.113E+00 CAL/MOLE/DEG
S  =  4.956E+01 CAL/MOLE/DEG
PQ =      33.469

RESULTS FOR THE REACTION:
         1.0 CL2 =    2.0 CL

DELTAH =   6.103E+04 CAL
DELTAS =   2.937E+01 CAL/DEG
DELTAG =   2.288E+03 CAL
AT T   =     2000.00 DEG

LOGKEQ =   -0.250

KEQ    =   5.622E -1

ENTER TEMPERATURE

T=-1

CPU TIME: 1.29  ELAPSED TIME: 4:56.60
NO EXECUTION ERRORS DETECTED

EXIT
```

Fig. 5. A continuation of Fig. 4. The equilibrium constant is generated for T = 1000° K and T = 2000° K. Execution is terminated by inputting T = -1.

eigenvalues, charge densities, and bond orders for bonded atom pairs. In addition, a complete output list is generated off-line. The program makes use of the SSP matrix diagonalization subroutine EIGEN. In addition, the research level program, NEMO (for nonempirical molecular orbitals [7] is set up for interactive teletype execution. Here, too, a selection of output is typed on teletype, and the complete output list is generated off-line. The author has made use of these two programs and DIATH2 in his quantum theory course.

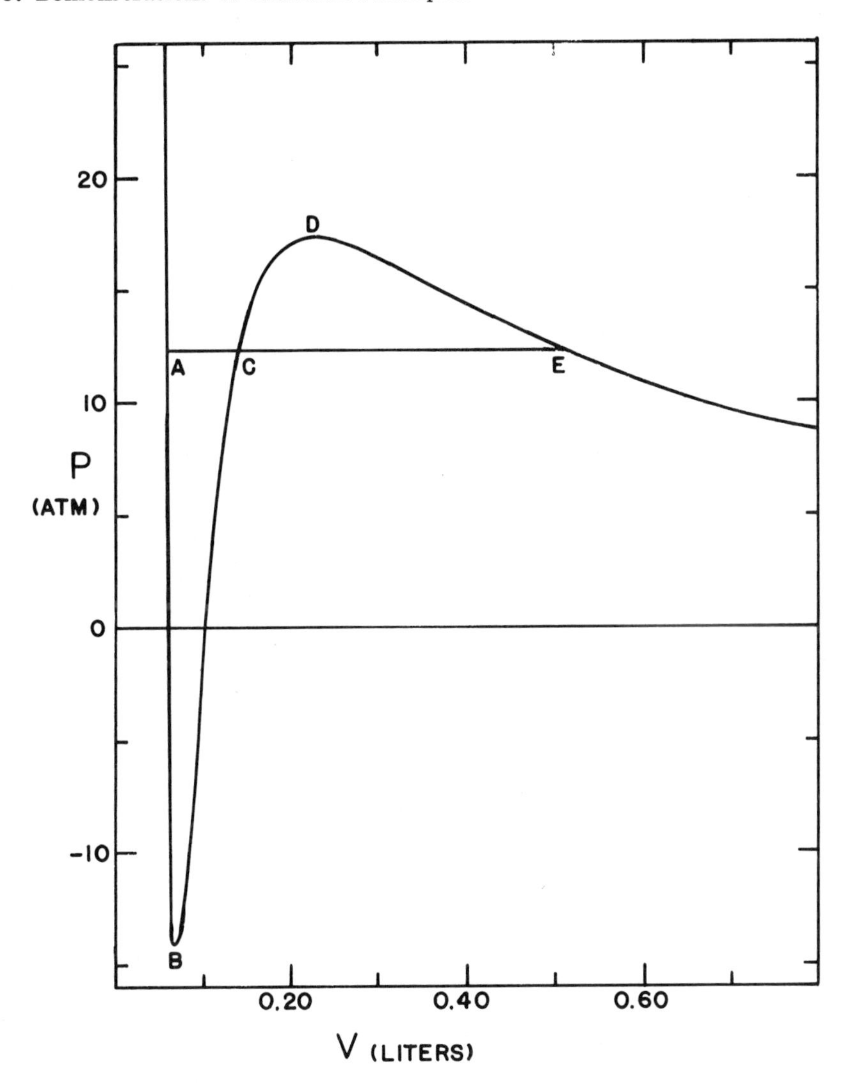


Fig. 6. Van der Waals isotherm for $N_2$ at T = 100 K. The value P = -14 atm at V = 0.08 l is genuine.

## 2. Thermodynamics

One of the author's students has just completed a program, GASLAW, which computes the ideal and van der Waals pressures of gases given T, V, and N as input. A table of parameters a and b for common substances is stored in the program, but the user can supply his own values if he wishes. This

```
R BASIC


READY, FOR HELP TYPE HELP.
OLD
OLD FILE NAME--VDW2

READY
RUN

VDW2            19:23           08-FEB-73



THIS PROGRAM COMPUTES VAPOR PRESSURES  FROM
VAN DER WAALS ISOTHERMS.
INPUT A(HE=0.03415, N2=1.346, H2O=5.468) ?1.346
AND B(HE=0.02371, N2=0.03852, H2O=0.03052) ?0.03852
ENTER T IN KELVIN ?50
VAPOR PRESSURE AT 50 IS 3.18099E-2 ATMOSPHERES
ENTER T IN KELVIN ?60
VAPOR PRESSURE AT 60 IS 0.659614 ATMOSPHERES
ENTER T IN KELVIN ?70
VAPOR PRESSURE AT 70 IS 1.82901 ATMOSPHERES
ENTER T IN KELVIN ?80
VAPOR PRESSURE AT 80 IS 3.98562 ATMOSPHERES
ENTER T IN KELVIN ?90
VAPOR PRESSURE AT 90 IS 7.40996 ATMOSPHERES
ENTER T IN KELVIN ?100
VAPOR PRESSURE AT 100 IS 12.3355 ATMOSPHERES
ENTER T IN KELVIN ?110
VAPOR PRESSURE AT 110 IS 18.9511 ATMOSPHERES
ENTER T IN KELVIN ?120
VAPOR PRESSURE AT 120 IS 27.4082 ATMOSPHERES
ENTER T IN KELVIN ?130
ERROR NO CHANGE
ENTER T IN KELVIN ?140
 140 IS ABOVE THE CRITICAL TEMPERATURE
ENTER T IN KELVIN ?-1
INPUT A(HE=0.03415, N2=1.346, H2O=5.468) ?-1


TIME:  49.65 SECS.

READY
BYE
```

Fig. 7. Sample input to and output from VDW2. The vapor pressure of $N_2$ is calculated at 10-degree intervals from T = 50° K to T = 140° K. The actual critical point data from CRC tables are $T_c$ = 126° K and $P_c$ = 33.5 atm.

program can be used to provide a quantitative basis for discussions of deviations from the ideal gas law.

She has also nearly completed a program, FEF, which tests the validity of the approximation

$$\Delta G = \Delta H^\circ + T\Delta S^\circ, \tag{14}$$

where $\Delta G$ is the free energy of a chemical reaction at an arbitrary temperature T, and $\Delta H^\circ$ and $\Delta S^\circ$ are the standard (T = 298° K) changes in enthalpy and entropy, respectively. The test is made by comparison with the more accurate

$$\Delta G = \Delta H(T) + T\Delta S(T), \tag{15}$$

where heat capacity data are used to evaluate $\Delta H(T)$ and $\Delta S(T)$ explicitly for the given temperature T. A table of heat capacity coefficients is stored in the program. Some work is still needed on these programs to make them useful for demonstration.

## ACKNOWLEDGMENTS

To Dr. Norman E. Sondak and the Worcester Area College Computation Center for furnishing computer time and advice on specific programming problems.

To the following WPI undergraduates for assisting on this project: David Lyons, Janet Merrill, Benjamin Thompson, and Christopher Williams.

## REFERENCES

1. (a) W. Heitler and F. London, Z. Physik, 44, 455 (1927). (b) Y. Sugiura, Z. Physik, 45, 484 (1927). (c) S. C. Wang, Phys. Rev., 31, 579 (1928).
2. (a) L. Pauling, Chem. Rev., 5, 173 (1928). (b) B. N. Finkelstein and G. E. Horowitz, Z. Physik, 48, 118 (1928).
3. C. A. Coulson, Trans. Faraday Soc., 33, 1479 (1937).
4. S. Weinbaum, J. Chem. Phys., 1, 593 (1933).
5. R. C. Plumb, personal communication. Another approach to the Boltzmann distribution is discussed in G. C. Pimentel and R. D. Spratley, Understanding Chemical Thermodynamics, Holden-Day, San Francisco, 1969. However, Pimentel and Spratley assume distinguishable particles and thus include a greater number of system states with multiply occupied particle states than are considered here.

6. H. B. Callen, Thermodynamics, Wiley, New York, 1960, pp. 146-154.
7. M. D. Newton, F. B. Boer, and W. N. Lipscomb, J. Am. Chem. Soc., 88, 2353 (1966).

1 - 49

158 - 173

189 - 完.

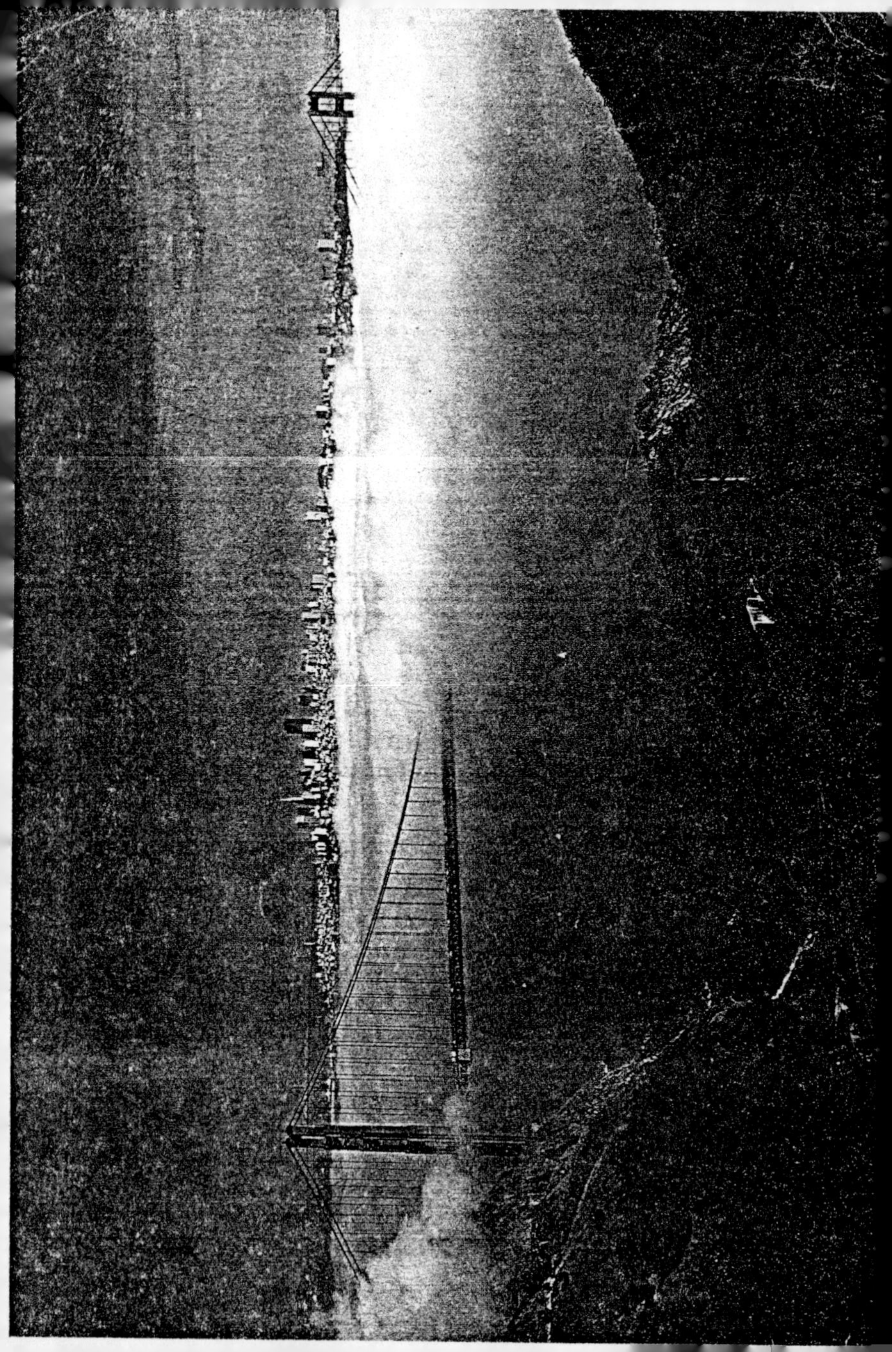

Chapter 6

# COMPUTER-SIMULATED UNKNOWNS

Clifford G. Venier and Manfred G. Reinecke

Department of Chemistry
Texas Christian University
Fort Worth, Texas 76129



## I. INTRODUCTION

Besides data processing, computers have been used in chemical education in three ways: instrument control, tutorial and drill presentation, and laboratory simulation. The first two mentioned areas are adequately covered in other chapters in this volume. It is our purpose to consider the area of computer-simulated laboratory experiments and, in particular, those involving the computer simulation of unknowns. Although there have been quite a few reports [1-14] concerning the use of simulated unknowns in the teaching of chemistry, details of the various systems are not always available. Therefore, when it is necessary or appropriate to use specific examples, we will take them from the system with which we are most familiar, namely, our own Armchair Unknowns program [1].

There are two kinds of problems in chemistry involving unknowns--those involving the determination of the identities of unknown substances, and those involving the determination of unknown quantities. The basic difference between these rests in the intellectual approach necessary to solve these differing problems. Although balancing a chemical equation or solving an equilibrium problem are skills that must be learned, such skills are not solutions of unknowns in the sense used in this chapter. While a limited number of standard methods can be used to solve equilibria or balance equations, the identification of unknown substances requires a logic that cannot be learned by rote. For the efficient determination of the nature of an unknown substance, contingency-conditioned feedback is an important part of the logical process. The kind of logical system used in solving an unknown is highly branched and not linear as is the logic of equation-balancing. Computer simulations of laboratory work which ask, "How much chromium is there in the iron ore sample?" do not represent unknowns as the word is used here, because the student simply follows a recipe, processes the numbers generated by the computer, and arrives at the correct answer. For the most part, then, we will discuss only the qualitative identification of unknown substances.

## II. RATIONALE FOR SIMULATED LABORATORY UNKNOWNS

Despite the many advances in its theoretical foundations, chemistry remains basically an experimental science. The emphasis placed on laboratory experience and training is therefore still as valid today as in Liebig's day. The laboratory unknown is a particularly useful pedagogical device not only for teaching chemistry but also for stimulating interest and providing an introduction to independent research on a small scale.

Part of the mystique of chemistry, part of everyone's interest in chemistry, stems from the fact that in the popular mind chemistry can determine the nature of unknown substances. To actually be able to participate in this process creates an interest among the students--an interest which hopefully will carry over to the other aspects of chemistry. The student who has depended on a cookbook recipe or upon careful observation of his neighbor's work for the successful completion of previous laboratory work is forced to attack a problem on his own. The ability to generate and interpret one's own data and hence the ability to apply basic principles to unique situations must be one of the most important prerequisites for a good scientist.

Computer simulations of laboratory experiments are desirable from several points of view. Although the computer cannot provide experience with the manipulation of equipment leading to the development of good laboratory "technique," it can assist the student in learning how to design the experiment and how to handle the data once it has been acquired. In addition, computer simulations of experiments allow the student to obtain experience quickly in the decision-making aspects of chemical experiments. It would be impossible to let each student run an experiment several different times in slightly different ways to get a feeling for how the results change with each of the possible variables. With computer simulation, however, each student can "run" the experiment many times in the course of a single afternoon, varying the conditions as he wishes, finding out for himself why some reagents are used in catalytic amounts, some in precise stoichiometric amounts, and others in great excess.

Computer simulation can also provide the student with the vicarious experience of experiments that are either too dangerous or too expensive for actual performance in the lab. Experiments that take a long time to complete can be "speeded up" on the computer. For example, ascertaining the results of a sodium fusion takes a few seconds with computer simulation, but hours in the laboratory.

Of all the things that can be done with simulation of experimental results on a computer, the one that uses the computer's special talents to the best advantage is the simulation of laboratory unknowns. Programmed texts and instructors can carry the student through the solution of linear problems, such as equation-balancing, on the premise that all of the specific equations encountered by the student in the future can be solved in a straightforward manner once he has learned the standard method. However, because of the highly branched nature of the solution of an unknown, it is impossible to give the student a single right way to solve every problem. To be sure, a start must be made somewhere, and the nature of the first few tests applied to the unknown may well be the same in each case. Nevertheless, after a few tests, each unknown becomes a unique problem, a problem which in fact has no single correct method of solution. Thus, each student is allowed to attack the problem in his own individual way, receiving as data only the information which he needs to solve the problem.

In addition, the computer simulation eliminates one of the major sources of frustration associated with qualitative analysis, the perplexing problem of wrong experimental results. Anyone who has taught, or taken, qualitative analysis knows that many incorrect identifications arise from technical rather than logical errors. In a computer simulation, a student need not be burdened with shortcomings in his laboratory technique while trying to build a logical framework on which to exercise his developing reasoning powers.

The computer will supply entirely reliable experimental results. The student need only supply the correct interpretation. Thus, the student can gain sufficient confidence in his reasoning powers so that when he begins actual laboratory work he will be able to decide if a particular inconsistency in his results is due to a mistaken inference or to an experimental error. At Texas Christian University we have found that students work faster and more accurately in the laboratory when qualitative organic analysis is preceded by exposure to a few simulated unknowns [1].

## III. CONSIDERATIONS IN CHOOSING OR PREPARING UNKNOWN SIMULATIONS

As in all matters of curriculum development, one must keep goals and resources firmly in mind while devising strategy. There are a number of questions that naturally arise in connection with the simulation of qualitative analysis.

### A. Computer Systems

Since we are not computer scientists and have experience with only a limited number of computer systems, we will not presume to recommend one particular computer system to the exclusion of all others. In reality, the individual chemistry teacher interested in incorporating computer-assisted instruction into his teaching will have little choice concerning the identity of the computer system he will use. There will probably be some sort of system already on campus, or arrangements already made to buy computer time from someone else. The teacher must make the best of what is available. However, in particular situations, the following discussions may be helpful.

One decision a novice may face is whether to use batch-processing for the generation of data or remote time-sharing terminals, which can present the student with information almost as soon as he asks a question. Again, the decision may be easy, since at some schools only batch-processing is available. However, more and more schools have available either typewriter-teletype terminals or cathode-ray tube (TV screen) terminals for CAI work. As usual, good arguments can be raised for either system. Direct interaction allows an almost conversational dialogue between computer and student. A sample from our own Armchair Unknowns program is as follows:

Student: Hi
Computer: Hi. Type in your student number, please.
Student: s39

Computer: Thank you, John. Which of your unknowns do you wish to work on?
Student: 2
Computer: OK. Unknown number 2 is a white solid. Which test or experiment do you wish to perform now?
Student: mp (melting point)
Computer: 31-33° C. Which test or experiment do you wish to perform now?
Student: sf (sodium fusion)
Computer: All tests negative. Which test or experiment do you wish to perform now?
Student: sl (solubility)
Computer: $H_2O$, no; NaOH, no; HCl, no; $H_2SO_4$, yes. Which test or experiment do you wish to perform now?
Student: off (signing off temporarily)
Computer: You are signed off now. Good luck with your analysis.

Student: Hi
Computer: Hi. Type in your student number, please.
Student: s39
Computer: Thank you, John. Which of your unknowns do you wish to work on?
Student: 2
Computer: OK. Unknown number 2 is a white solid. Which test or experiment do you wish to perform now?
Student: c7 (ceric nitrate reagent)
Computer: The solution remains yellow. Which test or experiment do you wish to perform now?
Student: c15a (hydroxylamine hydrochloride for aldehydes and ketones)
Computer: Orange solution turns red. Which test or experiment do you wish to perform now?
Student: c12 (fuchsin-aldehyde reagent)
Computer: Purple color develops. Which test or experiment do you wish to perform now?
Student: off (temporarily signing off)
Computer: You are now signed off. Good luck with your analysis.

Student: Hi
Computer: Hi. Type in your student number, please.
Student: s39
Computer: Thank you, John. Which of your unknowns do you wish to work on?
Student: 2
Computer: OK. Unknown number 2 is a white solid. Which test or experiment do you wish to perform now?

| | |
|---|---|
| Student: | d16 (semicarbazone) |
| Computer: | White solid formed, mp 215-218° C. Which test or experiment do you wish to perform now? |
| Student: | d49 (oxime) |
| Computer: | White solid formed, mp 90-92° C. Which test or experiment do you wish to perform now? |
| Student: | 22381 (o-methoxybenzaldehyde) |
| Computer: | 22381 is incorrect. Recheck your data. Which test or experiment do you wish to perform now? |
| Student: | bp (boiling point) |
| Computer: | Greater than 260° C. Which test or experiment do you wish to perform now? |
| Student: | 22342 (alpha-naphthaldehyde) |
| Computer: | Congratulations, 22342, alpha-naphthaldehyde is the correct answer. If you wish to rest on your laurels, type "off". If you wish to try another of your unknowns, type "index". |

(Each test or experiment and each unknown is provided with simple codes derived from Shriner, Fuson, and Curtin's text on qualitative organic analysis [15]. For an explanation of the codes, see [1]. The comments in parentheses are not actually part of the dialogue but are included here for the benefit of the reader.)

The direct interaction technique allows the student to work through an unknown in a very few minutes. He need only stop when he reaches the point where he must consult the literature before performing another experiment. Note that in the sample dialogue, the student pauses twice to consider the results he has just acquired; he always knows the answer to one experiment before performing the next. This means he will be able to utilize to the maximum the contingency-feedback nature of the reasoning which is so important to the proper handling of the identification of unknowns. The student in the above example did not run a boiling point as a preliminary test since his compound was a solid, but when a boiling point would serve to distinguish two possibilities, it was run.

Batch-processing, besides being less costly, more closely resembles the lab in one respect--with a typical turn-around time of 24 hours, some of the frustration and anxiety associated with the actual laboratory can be experienced. In addition, while the student awaits the result of one test, he must be considering his next step, keeping in mind that the results of the test may be negative as well as positive, and therefore learns twice as much chemistry.

Another technique for the presentation of data to the student involves the use of random-access carousel slide projectors. Wing [10] has described a prototype system involving 1500 slides for qualitative inorganic analysis.

In answer to his inquiries, the student is shown a picture of the result rather than being given a verbal description. This would have the advantage of familiarizing the student with precisely which yellow color represents a positive test. In addition, there would be no clues in the presentation concerning the nature of the observation one is looking for. In verbal presentations, one generally notes, "Gas evolved" or "Turns yellow". A picture presents the whole result of the experiment and not just the pertinent data. Thus, the student must learn the chemistry of the tests well enough to know which property of the system should be observed.

In qualitative organic analysis simulations which use spectral data, the use of slides of the actual spectra of the compound and some of its important derivatives, can likewise be stored and made available to the student upon request. In the situation where points are charged against the student's account for information, there is an advantage to having spectral data available only on request. In the usual simulation using spectral data, it is handed to the student as part of a set of mimeographed handouts concerning the use of the computer, or is presented in tabular form by the computer. The tabulation of spectral data, especially infrared data, requires a judgment on the part of the tabulator as to which lines are important and which are not. Pedagogically, it is better to allow the student to figure out for himself which parts of the spectrum are important.

Many different languages have been used for the programming of computer-simulated unknowns. COURSEWRITER I and II were developed by IBM specifically for CAI programming, but both are now in disfavor because they lack computational power. FORTRAN, BASIC, and APL have all been successfully applied to computer-simulated unknowns; these languages allow one to make numerical computations during the course of execution of the programs, thereby allowing the student wide latitude in places where he must supply numerical data.

## B. Strategic Concerns

Whether one chooses to use a program developed by someone else or chooses to develop his own computer-simulated unknown program, he needs to be aware of a number of considerations involving strategy.

It is important to consider the purpose to which a CAI unknown simulation is to be put before choosing a program or a system to use. There are two important general uses for which unknown simulation may be used. First, simulation is an excellent introduction to laboratory work. The logical process of the analysis is learned independently, before actual laboratory work begins. A second use of such simulations is to supplement courses which do not ordinarily treat qualitative analysis. In a rather short time, the systematics, logic, and some of the chemistry involved in analysis can be appreciated without the extensive lab work otherwise necessary. Qualitative inorganic analysis in freshman chemistry and qualitative organic analysis in nonmajor terminal courses and "short" (two-quarter or one-semester) service organic courses are enriching curricular supplements. Ordinarily, the simple programs with fewer tests and fewer possible unknowns are best to illustrate the nature of the identification process. More elaborate programs more closely resembling actual laboratory work are more appropriate to cases where simulation will be used as an introduction to a qualitative analysis laboratory.

Once the purpose is well in mind, one must decide whether the students will be expected to take more or less the same route toward the goal of identification or whether it is sounder to let each student discover for himself the most efficient way to identify an unknown. Generally, one would choose to use one of the tried and tested methodologies for solving unknowns, usually the $H_2S$ scheme for qualitative inorganic analysis and the solubility classification scheme for organic qualitative analysis. Unless there is quite a bit of time available and the students are bright enough to be able to work out their own mini-schemes, it is wise to provide some sort of underlying scheme on which the students can rely.

Next, one must select a textbook which presents the chosen scheme. For the student working at a computer terminal, such a textbook serves as a handy reference book to aid in the choice of tests and the interpretation of results. In addition, one may choose to limit the number of possible unknowns to those treated in the scheme presented by a particular textbook. In qualitative inorganic analysis this is the traditional approach, many metals being left out of the scheme in order to keep the laboratory work both simple and safe. Traditional qualitative organic analysis texts also severely limit the classes of compounds from which unknowns are chosen; within each class the actual possibilities are further limited to the well-known members. Since the purpose of the computer simulation is to familiarize the student with the logical processes involved in the identification of an unknown, we believe that a wisely limited set of representative organic compounds, such as those listed in traditional texts or the CRC's Handbook of Tables for Organic Compound Identification [16], serve the purpose well. With the whole of the literature from which to choose a compound, it is little wonder that Sherman [5] reports that students become frustrated when they have run vitrually every test in their battery and there are still two or more compounds which are consistent with all of the data. Just as a data bank containing every conceivable reagent and test cannot be constructed, neither can every known compound be used as an unknown.

An important decision for the instructor arises over the question of whether to provide spectral data or not. Although spectral data are extremely important in the real world for the identification of substances, part of the raison d'être for qualitative analysis is the chemistry learned in the process of identification. Students find, as have chemists in general, that spectra are the easiest and surest method of identification. Often they will work entirely from spectral data, if it is provided. If there are spectral indexes handy (Aldrich's infrared library [17], Varian NMR catalogues [18], or the Sadtler indexes [19], students will opt for spectral identity as the only means of really being sure of an identification, and thereby lose the opportunity to learn a substantial amount of "wet" chemistry. We have found that with a limited set of possible unknowns, students learn more chemistry without spectra. The interpretation of spectra, while important, is probably better treated separately.

With respect to the actual dialogue between the student and the computer, one can choose a very natural exchange of English words between the two, either by key-work recognition [8] or by the more sophisticated syntactical analysis of the PLATO IV system (see the chapter by S. G. Smith in this volume), or one can use very short coded responses representing the limited set of tests or reagents that the student can use [1,3]. The natural responses are surely esthetically more pleasing, but the coded responses have important advantages for the typical instructor and student. Coded responses of two to five letters and numbers are easy to program and convey more precise information than the usual English sentence. Short responses reduce the chance for input errors, save time, and do not penalize the student who has never learned to type.

## C. Evaluation

Motivation and evaluation are inextricably intertwined. The reward of points for a good performance on a test serves both to motivate the student and to provide a measure for the instructor's evaluation of the student's work. If the student is learning a skill that will help him later in the course of his laboratory work, he may well be motivated simply by his desire to be able to handle the new laboratory situation as easily as possible. In cases where a more positive motivation or an actual evaluation is desired, one can assign point values to the various tests and allow the student to solve the unknown while using as few of his allotted points as he can (see [3], [4], and [5], for examples). This tends to maximize efficiency; the student will quickly learn to perform the tests most likely to give him the most information. A variant of this type of evaluation involves conversion of the unknown simulation into an economic game. Instead of points, the student is charged dollars for tests on a predetermined schedule [4, 5].

The main objection to the foregoing type of evaluation is that only a running total of points or dollars expended is kept for evaluation by the instructor. Since one of the prime purposes of having unknowns in the laboratory, and hence in using unknown simulation, is to develop in the student the ability to attack the problem logically, it is important that the instructor assure himself that the students are not guessing without sufficient data in order to minimize their point expenditure. For this reason, we recommend that regardless of the method chosen for assigning a grade, some sort of condensed performance record be made, detailing the tests used and the order in which they were performed. The instructor can then evaluate the method by which the student has attacked the problem. We have found that going over these performance records individually with the students is a good instructional aid. Alternatively, the simulation could be programmed in such a way that the student must perform a minimum set of tests before the computer would accept and evaluate even the correct answer. For example, if a student has the single unknown cation lead in a qualitative inorganic analysis simulation, and he immediately treats the solution with $H_2S$ in HCl, and observing a black precipitate, informs the computer that his unknown is lead, he surely has not correctly identified the unknown even though the answer is right. If only correct answers and points were totalled he would appear to be a genius, when just the opposite is true. One can program the computer in such cases to inform the student that he has insufficient data upon which to make his judgment [11], and either let him continue or comment concerning which factor he has failed to consider.

## IV. EXTANT PROGRAMS FOR UNKNOWN SIMULATIONS

In the sections below, we summarize some of the programs for unknowns that have come to our attention. Many chemists write and use educational materials utilizing computers and never report their results to other chemists. When results are reported, they are often contained only in the proceedings of one or another meeting. Even when they are reported in the open literature, details of the programs and systems are often distressingly vague. Thus, we apologize to the many workers in the field whose results will not be summarized in this chapter. For most of the programs below, the reader need simply write to the author of the program for a further description of copies or the programs.

In addition, if one is developing his own programs, the description of how others have solved similar problems will be useful.

### A. Qualitative Inorganic Analysis

Qualitative inorganic analysis programs generally follow the traditional laboratory practice using the $H_2S$ scheme. Sherman [4] uses a batch-process on a CDC 3300 computer programmed in the FORTRAN IV language. The student is charged a fee for each analysis and tries to correctly identify the cations while spending the least amount of money. The instructor gets a print-out of the student responses in order to supervise the learning experience. The author claims that his program is easily transportable to any 25K computer which will accept FORTRAN IV. The program is available from the North Carolina Educational Computing Service [20] or from the Eastern Michigan University Center for the Exchange of Chemistry Computer Programs [21].

Wing [10] described the use of random-access carousel slide pictures as responses to student requests for information about their unknown. It is programmed for IBM computers.

The University of Texas group has developed and carefully evaluated a qualitative inorganic analysis simulation [8, 12, 13]. These programs were originally written in the COURSEWRITER languages for IBM 1440 and IBM 1500 systems. This group is now using a CDC 6400 system; programs compatible with that system are available as well. The programs can be used with either CRT or teletypewriter direct-interaction terminals.

## B. Qualitiative Organic Analysis

Qualitative organic analysis programs are generally based on initial classification of unknowns by solubility, in that solubility data are always provided. The student, of course, may choose to ignore the traditional approach and wade into the unknown randomly. One of the beauties of the computer simulation of the laboratory exercise is that the student can discover for himself that random-testing is counterproductive.

Venier and Reinecke [1] have described a system based on Shriner, Fuson, and Curtin's textbook [15] which utilizes that text's 3000 possible unknowns as a limited set of unknowns. Versions of the program in COURSEWRITER II for the IBM 1500 system and APL for the IBM 360 series have been developed. The COURSEWRITER program uses CRT direct-interaction terminals; the APL version uses teletypewriters. This system gives the student a choice of about 120 tests. Simple codes, usually of only a few characters, are used for student responses.

The system written by Crain [3] is quite similar to that of Venier and Reinecke. His program is written in TIME-SHARING FORTRAN for a Honeywell-635 computer, but he states that he has programmed it with easy conversion to FORTRAN IV in mind. Again, a limited set of possible unknowns is used, those in Shriner, Fuson and Curtin [15] or in the CRC's Handbook of Tables for Organic Compound Identification. Fifty tests can be simulated.

The University of Texas group has a qualitative organic analysis program [6, 7] available in COURSEWRITER and BASIC versions which do not limit the number of possible unknowns.

Smith [2] at the University of Illinois has developed a qualitative organic analysis system using the PLATO system. Although his system will not be directly applicable for others, the PLATO system will someday have a large number of users at remote terminals over a wide area (see S. G. Smith's chapter in this volume).

The North Carolina Educational Computing Service [20] has available a qualitative organic analysis program entitled IDGAME, written by Hornack [5]. This program served as a model for Sherman's qualitative inorganic analysis program and possesses about the same features, including batch-processing and a financial system of score-keeping. It is programmed in PL-1 and has the important advantage that it is available in completely documented form from NCECS.

### C. Other Programs of Interest

The key element of the previously described qualitative analysis programs is that they are student-managed. The student is not given a complete set of data and asked to interpret it. He must understand the material on a higher plane. He must be able to decide at each point in his analysis which piece of information of the many from which he can choose would be the most valuable to acquire. The programs described below put a premium on and reinforce independence and understanding in much the same way that the qualitative analysis programs do. In each, the student must make a positive commitment in selecting for himself the exact conditions for experiments. These programs, then, have much in common with qualitative unknowns and are therefore included here.

One interesting use of the computer's ability to provide data only upon request is illustrated by a kinetics simulation program written by H. W. Smith of Earlham College [22]. The object for the student is to determine the rate law for a reacting system by requesting concentration vs time data under a set of conditions specified by the student. The student is charged money (points) both for the number of experiments to be performed and the accuracy to which he wishes to determine the concentrations and time. The student chooses initial conditions and then asks for data at nominal times. The computer generates concentrations within the deviations previously determined by the student, the variances being randomly incorporated into the calculations by the computer. Thus, the student must learn not only to process data concerning kinetics, but also must learn to choose the right experiments in much the same way one learns this skill in the laboratory.

Titrations may also be simulated in a student-managed format. For example, the equivalent weight of an unknown acid or the percent composition of a quantitative analysis unknown can be determined with the student specifying all of the reagents and conditions and even running the titration a point at a time by specifying the amount of titrant to be added at each new point. The University of Texas group [12, 13] has a number of such simulations. They are direct interaction programs written in COURSEWRITER or BASIC.

## V. CONCLUSIONS

The state of the art for the production of computer-simulated unknowns is sufficiently well developed both in theory and in practice that anyone with a computer, knowledge of programming and some good ideas can produce unknowns to aid in his instructional program. In fact, only the good ideas are necessary if the computer center staff or a colleague will assist in the programming. In addition, such places as the Eastern Michigan University Center for the Exchange of Chemistry Computer Programs [21] and the North Carolina Educational Computing Service [20], plus individual authors, have previously prepared and tested materials available just for the asking. Thus, anyone can have the use of simulated unknowns very quickly.

Simulated unknowns have the advantages that they (1) allow one to perform experiments in a fraction of the time usually used in the laboratory, (2) divorce the technical laboratory manipulations from the intellectual content of analysis so that latter can be mastered independently, and (3) allow a variety of experiments to be experienced that would otherwise be unavailable due to cost, danger, or abnormal reaction time. Lastly, they are loads of fun.

## REFERENCES


1. C. G. Venier and M. G. Reinecke, J. Chem. Ed., 49, 541 (1972).
2. S. G. Smith, J. Chem. Ed., 47, 608 (1970).
3. R. Crain, University of Kansas, private communication, 1972.
4. L. Sherman, The Science Teacher, in press.
5. F. Hornack, Proceedings of the Conference on Computers in the Undergraduate Curricula, Dartmouth College, Hanover, New Hampshire, June, 1971, p. 359-362; L. Sherman, Proceedings of the Conference on Computers in Chemical Education and Research, Northern Illinois University, DeKalb, Illinois, July, 1971, pp. 6-13 to 6-22.
6. G. H. Culp and S. J. Castleberry, Sci. Ed., 55, 423 (1971).
7. G. H. Culp and J. J. Lagowski, J. Res. Sci. Teach., 8, 357 (1971).

8. T. T. Hollen, Jr., C. V. Bunderson, and J. M. Denham, Sci. Ed., 55, 131 (1971).
9. V. S. Darnowski, The Science Teacher, 35, 22 (Jan., 1968).
10. R. L. Wing, The Science Teacher, 35, 41 (May, 1968).
11. J. A. Swets and W. Feurzeig, Science, 150, 572 (1965).
12. S. J. Castleberry, E. J. Montague, and J. J. Lagowski, J. Res. Sci. Teach., 7, 197 (1970).
13. S. Castleberry and J. J. Lagowski, J. Chem. Ed., 47, 91 (1970).
14. L. B. Rodewald, G. H. Culp, and J. J. Lagowski, J. Chem. Ed., 47, 134 (1970).
15. R. L. Shriner, R. C. Fuson, and D. Y. Curtin, The Systematic Identification of Organic Compounds, 5th ed., Wiley, New York, 1964.
16. Z. Rappoport, Handbook of Tables for Organic Compound Identification, 3rd ed., Chemical Rubber Co., Cleveland, Ohio, 1967.
17. C. J. Pouchert, The Aldrich Library of Infrared Spectra, Aldrich Chemical Co., Milwaukee, Wisconsin, 1970.
18. N. S. Bhacca et al., NMR Spectra Catalogue, Vols. 1 and 2, Varian Associates, Palo Alto, California, 1963.
19. The Sadtler Standard Spectra, multivolumes, Sadtler Research Laboratories, Inc., Philadelphia, Pennsylvania.
20. North Carolina Educational Computing Service, Box 12175, Research Triangle Park, North Carolina 27709.
21. EMU-CECCP, Chemistry Department, Eastern Michigan University, Ypsilanti, Michigan 48197.
22. H. W. Smith, Proceedings of the Conference on Computers in Chemical Education and Research, Northern Illinois University, DeKalb, Illinois, July, 1971, pp. 6-22 to 6-29.

Chapter 7

# APPLICATION OF CANNED COMPUTER PROGRAMS TO THE UNDERGRADUATE CHEMICAL CURRICULUM

Larry R. Sherman

Department of Chemistry
North Carolina Agricultural
and Technical State University
Greensboro, North Carolina



## I. INTRODUCTION

This chapter is not an attempt to be an all-inclusive discussion of the use and operation of canned programs, but rather is an attempt to discuss some of the problems, successes, and failures of using black-box computer technology in the chemistry curriculum of a small college. Much of the work is strongly oriented toward analytical chemistry because this is the author's field of greatest involvement. Many good systems and ideas have been neglected, some because of oversight, but more because the documentation is poor and the author has no personal knowledge of success.

In this chapter a "canned" program refers to one available in a card deck; it may or may not be disk- or tape-mounted, but does not refer to interactive-type programs. The operation of a keypunch is the only knowledge required of the students.

Too often instructors see Computer-Assisted Instruction (CAI) as a gimmick and fail to use the available material because they have not kept abreast of the modern curriculum developments. The computer should be put on par with overhead projectors, films, and other audio-visual aids. Too often the gimmick-suspecting instructors rationalize that a computer reduces the individuality of the students; the direct opposite is being proven [1].

There are also instructors who think the computer is "God's gift" to the teacher and attempt to do all their teaching with it. This kind of approach can produce students who are incompetent chemists, mathematicians, etc., when thrown into situations with no available computer facilities [2]. The computer, and especially the use of canned computer programs, lies somewhere between the two extremes. They should supplement the curriculum where the work becomes too tedious or the equipment is unavailable.

Students' reception of CAI seem to be directly related to their grade-point average. Those students who earn A or B usually produce good results; those who earn C or D usually produce poor results. The CAI cannot be blamed for the lack of motivation, as the same students who produce poor grades usually fail to do homework assignments or turn in reports on time.

In the past few years a large number of computer programs have been written to help teach chemistry. In the Journal of Chemical Education, the largest journal devoted to the field of chemical education, approximately 116 papers have been published on teaching computer programming to chemists and applying computer programs and programming to chemical problems (between 1965 and 1971). Even though this constitutes about five percent of the papers in the Journal since 1965, this number is small compared to the amount of work being done in the field. More papers have been published in the proceedings of various conferences [3], and a few have found their way into more prestigious journals [4]. Yet for every paper that is published in this area, approximately five workable ideas are not published either because the authors do not wish to take the time to document their work or because the articles are refused by journals [5]. Despite these handicaps, chemical educators have disseminated their work with such disregard for personal gain that their unselfish devotion to furthering the chemical field should be honored by those in the more cutthroat research areas where every set of data is strictly guarded until published.

Attempts have been made to create centers for distribution of computer-oriented chemical education information; however, none of these as yet have the esteem, acceptance, or smooth administration of the Quantum Chemistry Exchange Center at the University of Indiana [6]. Until such a center is in operation on a national level, the educator will have to trust to hearsay and personal contact, and suffer the frustration of a lack of standards.

If the educator desires computer ideas, he can easily acquire 150 to 200 programs. However, about half of these can never be adapted to the local computer system; they just will not run. To shrug one's shoulders and state that this is a transportability problem is to circumvent the problem and not attempt a solution. Lack of transportability discourages the beginner in his attempt to give his students the best that modern technology can offer, frustrates the novice, and greatly delays the skilled programmer. That educational programs are written in a large number of computer languages is not the major problem in transportability; while the reduction of languages to two or three would greatly help the situation, most of the trouble lies in the documentation of the programs themselves. If an educational program is to be useful, it must be internally and externally documented. More often than not, either or both are missing; the educator finds it difficult to determine exactly what principles are covered in the work, what data are needed, and what the final output will show. If such documentation is available most programs can be transported from one computer to the other. Often minor modifications need to be made to handle individual quirks of a particular computer or instructor, but any good programmer or system analyst can make the necessary changes in an afternoon. Sometimes it is easier to take the basic ideas and write a new program to solve the problem. The latter case often entails translating one language into another. Any proficient programmer who understands systems inputs and outputs and internal movement in a computer can translate any of the major scientific languages into another.

Two basic types of canned programs are used by the educator: (1) number-crunching programs, where the students perform the chemical experiment or preliminary calculations and then insert the data in the program which performs the arithmetic, and (2) programs that simulate chemical instruments or experiments and give data that look like that obtained in an actual laboratory situation. In the latter case the students are often allowed to enter variables into the operation which would be totally unacceptable in a real laboratory since they could burn out components of the instruments or cause unacceptable laboratory results (e.g., fire). In this case students can be taught the operation of a complicated instrument without the possibliity of damaging it. However, a computer-simulated chemical experiment is not a substitute for the actual laboratory work; it can only be a supplement. It is of little use for a student to be able to calculate force constants, transfer coefficients, etc., if he does not have any idea which instrument will give him the needed data [7].

## II. NUMBER-CRUNCHING PROGRAMS

This type of canned program should be the most widely used computer-assisted instruction, but apparently is the least used by students or instructors in chemistry. This is not true in Physics and Mathematics. Books have been written to teach these subjects, where all the number-crunching is performed by canned programs. These programs are either supplied by the instructor or provided in such a way that any student can make the minor changes needed for his university computer and punch the card deck. To the author's knowledge, there are no books available in chemistry to do this work, although some are planned. The lack of use may be due to the assumption of many instructors that "if a student needs help in computing data or solving long problems, let him write his own program." This rationale has a great deal of merit, but few students, including many graduate students, have the expertise to write the type of program needed to work most chemistry problems. Furthermore, since more time is usually involved in debugging a program than in writing it, most students who do have the writing knowledge prefer to work the problems by longhand or with a slide rule. One can hardly blame a student for this type of approach, but an instructor who is concerned about the amount of work his students perform and who is interested in curriculum development should look at the total number of man-hours spent by 10 to 50 students in performing arithmetic that teaches neither chemistry nor mathematics. An individual student may not feel that three to five hours is worth the time in writing a number-crunching program, but an instructor should feel that ten hours of his time is worth 50 or 100 hours of student time; he should be willing to prepare the programs and make them available in such a manner that a student would only need to provide the data cards.

All canned programs should be used a black boxes. This allows the instructor to use programming and mathematics too sophisticated for the average student to comprehend and allows the use of standard programs written to handle more than the student's need. It allows programs which were written for number-crunching of research data to be used for trivial student work. Furthermore, student programs are written to solve one specific laboratory or lecture problem and are useless after their initial use.

## A. Linear Regression

Probably the most useful and universally accepted number-crunching program is a simple least-squares fit of data to a straight line. Every instructor who has taught either physical chemistry or instrumental analysis feels that his students need a better method of drawing straight lines than sighting along a ruler [8]. Since these courses are taught to upper classmen, the students are often asked to write a computer program to perform a least-squares fit on their data. The programming is usually good, but the chemistry can often be a disaster. Furthermore, even a computer-calculated straight line is only as good as the data used to calculate it; and if the work is accomplished without the statistics on the points, the end product is no better than that obtained with a ruler on a piece of graph paper. A method has been suggested to correct for some of the errors normally deleted [9], but this addition to the programming prohibits the ordinary student from writing the program.

A large number of canned programs and subroutines exist in this area [e.g., F4REG (Philco system), LINREG (author), LEASTSQ (NCECS)]. The black-box approach places the emphasis on the chemistry involved in preparing the data points, and the best straight line through the points can be used in evaluating the data without the tedious calculations needed for long-hand computations. The statistics for the data can also be provided by the program, and the students can observe the relative errors encountered in the experiment. An enterprising instructor can use this type of canned program in his general chemistry course for experiments concerning first-order kinetics [10].

A canned least-squares fit of data emphasizes the limitation of an instrument's sensitivity, reproducibility, and accuracy. It also tends to reduce the "godlike" awe some students assign to the numbers obtained from an electronic apparatus.

## B. Polynomial Solution

Chemists tend to do a much poorer job of solving the mathematics associated with their work than physicists and as such make less use of the sophisticated methods of obtaining answers to their work. This is especially true in the solution of complicated pH or complex ion concentration problems. They often have no trouble setting up the problem and rapidly reducing the non-linear simultaneous equations to a complex polynomial. At this point they usually make assumptions, often invalid, to simplify the equation, or resort to tedious numerical analysis techniques to obtain an approximate solution [11]. By using a canned computer program the polynomial equation can be solved by the computer and the chemist receives an answer that is as exact as the data used in the problem [12].

Quantitative analysis students dread being asked to solve exact pH or complex ion problems. The algebra in solving the simultaneous equations is difficult, but the worst part of the problem is solving the polynomial equation. Students rarely make a good assumption about which terms can be ignored in the final solution. This is usually due to their inexperience with actual situations. Thus, they feel they must go through the tedious numerical analysis to avoid "losing points" for invalid assumptions. A number of short cuts [13, 14] have been published about this type of problem, but the best solution involves the use of a computer program to solve polynomial equations after the students have performed the algebra.

### C. Acid-Base Titrations

Most instructors who use CAI eventually use some type of acid-base titration program. PALS [13] lists 13 variations on this theme; more are available [14, 15]. Most instructors of quantitative analysis realize that there is little merit in having the students perform a point-by-point calculation on more than one titration curve. It is also impractical to make students perform more than one strong acid-strong base and more than one strong base-weak acid experiment in the laboratory. The rest of the acid-base titration theory must be presented in lecture. No matter how well it is presented, about half of the students have little comprehension as to why the curves depicted in the textbook are obtained.

A canned computer program which accepts the $pK_a$ or $pK_b$ and other typical titration data allows the student to repeatedly try the principles and observe what happens as one modifies the titration parameters. HA is a program which has been used in several locations with considerable success [12, 16].

To illustrate the problem, the instructor can assign a homework problem that asks the students to plot and analyze five to ten titration curves with $pK_a$ of 1 to 9. As the students plot the various curves, they usually question the shape. Almost all of the students [12] using this method have understood the hydroxide leveling effect in a weak acid-strong base titration and quickly grasp the relationship between curve shape and hydrolysis.

Two problems are inherent in these canned programs: (1) Most titration programs solve a quadratic equation. Students usually forget that water will contribute a significant hydronium or hydroxide ion concentration to a solution that is very dilute or in which the $pK_a$ value is very high. This means that inept use of the program will give the student a distorted picture of a real titration. (2) Many instructors become excited about the results obtained from their students and add a plotting routine to the calculations. In doing so they remove much of the thinking related to the problem and tend to destroy the success which they observed earlier.

Although a nonaqueous titration is quite similar to aqueous titration, there does not seem to be any documentation on this type of titration to show the extrapolation of anion leveling effect of a solvent other than water.

## D. Electron-in-a-Box

The calculation of the energies of an electron-in-a-box is often looked upon as an academic exercise to introduce students to quantum mechanics. Some instructors assign this as a routine homework problem, and it is found as a homework problem in some textbooks. Several programs are available for doing the mathematics in a number of languages and in many modes of CAI (e.g., canned, conversational, or as subroutines). Many times the programs are used by the students to check their calculations. The program then degenerates to using the computer as a calculator.

These programs can be used to supplement laboratory work. Several experiments have been published to show the relationship of absorption bands of conjugated molecules and the bands predicted by quantum mechanics [17] if the correct assumptions are made for the energy and bond length. Most undergraduates cannot perform these computations without taking a course in quantum mechanics. Many instructors circumvent this problem by using formulas [18] which are more empirical than theoretical. The formulas agree quite well with the observed phenomena but tend to give the students the idea that theoretical calculations really do not work.

The program BOX [19] is a one-electron program which asks the student to enter only the length of the electron box (in angstroms) and the energy of the box walls (in electron volts). After a simple introduction to quantum mechanics in the classroom, most students can make reasonable guesses about these values if they take the time to draw the structure of the molecule and realize that the height of the box walls depends upon the ionization energy of the molecule. The students who enter the correct parameters receive the energy for the first, second, or third energy level which agrees with the energy of the absorption band observed in the laboratory. Normally less than 25% of the students will obtain answers anywhere near the observed energy. If they are required to explain their answers, they often realize that the length of the box was in error or that the energy of the walls was very poor in comparison to the known energies for electronic vibrations. After a discussion and review of the experiment, the students are asked to perform another calculation based on their new knowledge. The results are always good.

One of the advantages of a canned program like this is that the students can immediately see the usefulness of quantum mechanical calculations. Since the tedious computation is removed, they are willing to repeat calculations until they obtain answers which agree with the experimental results.

The use of the canned program brings the theoretical lectures into the laboratory, and even allows the students to see how they might identify compounds and other physical chemical parameters in the future. The disadvantages are found with the poorer students--they are often lazy and cannot or will not make reasonable assumptions; they are often unwilling to spend any time looking up data in the literature; and after two or three bad attempts at obtaining the data needed, they resort to copying the data of one of the better students.

## E. Infrared Searching

Probably the most useful canned program ever provided to students and faculty is the Infrared Spectral Information System (ISIS) [20]. ISIS is a master program for searching and identifying over 100, 000 compounds from their IR spectrum. Other search programs are available such as IR and IRSIRCH [21], but no other program covers the eight files that ISIS embodies (API, Sadler, NRC-NSB, ASTM sponsored, Doc. of Mol. Spect., Coblentz Soc., MCA, and IR Committee of Japan).

To anyone who has performed any research on unknown organic compounds, the usefulness of ISIS is immediately evident, but its use in the chemistry classroom has be infrequent. In most upper-level organic and instrumental analysis classes, students are given instruction in IR identification and some routine operation of simpler IR instruments. To simplify identification of compounds the students are usually provided with pure organic substances and asked to identify the compounds; sometimes, to help the students, the molecular weight of the compound is provided. Most students spend a week searching reference books on IR and identify the most obvious bands, the CO, OH, aromatics, NH, etc., but rarely come up with the right answer. Their textbook tells them that an IR itself is not a means of positive identification, so if they obtain an improbable structure, they quote their textbook. Some enterprising students acquire the instruction's code sheet or find access to the store room, and equipped with the molecular weight and most characteristic bands, search the shelves until they find a compound which looks correct. They may even take a melting point and turn in an answer. Students who have shown such enterprise are usually rewarded for their hard work with an A for identifying the compound. However, every instructor knows that this is not a realistic method of identifying compounds; rarely does an analyst have a fixed supply of compounds (as the instructor) from which to choose his answer, so he must resort to other methods of identification. Furthermore, those students who are not resourceful or are unwilling to spend the added hours become frustrated and will often make wild guesses even after they have successfully identified the most important functional groups. The problem in IR identification is that very few people, even those who consider themselves skilled in identification, can adequately identify and interpret the fingerprint regions of the spectra.

Since most chemists only use IR identification on rare occasions, it is actually a waste of time teaching every student the shift in bands which give characteristics to the fingerprint area. However, it is important that all chemistry students be able to quickly identify the most common functional groups; therefore, ISIS can be used as a supplement to a course where IR identification is part of the training. A great deal of success has been achieved [12] if the introduction of the ISIS search system is made after the students have made their own attempt at identification of a compound. With the computer the students can obtain their results in a few minutes. Care must be exercised in cautioning the students not to be over-zealous. They often want to code everything into the program and thus eliminate their compound. They also try to be accurate and include numbers which are not significant figures; since the numbers are truncated rather than rounded off, the students obtain erroneous results or none at all. Since ISIS returns the most important bands searched as well as the compound names, the students can check their work in handling the computer program. Success is usually achieved after two or three tries.

The biggest disadvantages of ISIS is that it is written in assembler language, which means that the punch cards must be more carefully prepared than with other canned programs. Furthermore, the quantity of data needed for each search makes it extremely difficult to prepare a set of punch cards without error. However, the success of the program does compensate for these disadvantages.

After learning about ISIS some students use it to identify their qualitative organic unknowns. Since the students identify their unknowns without learning many of the procedures involved in identification, ISIS can defeat the purpose of the qualitative organic course. To counter this, some instructors forbid their students to use a computer search method, but this denies the student a useful tool he will need in the future. The ideal situation is a compromise between an open and closed ISIS file, but as yet no one has documented a good compromise.

ISIS has been also used with an honor class of physical science students [22] to identify pesticides in an ecology game. Limited success was achieved due to the small number of students involved in the game. The ecology game opens the use of the program to the unskilled students who know little chemistry, who are not likely to learn very much chemistry, but who want to be in the foreground of current fads.

The ISIS file has been checked for use in structural studies [20] but no documentation on this use is currently available.

## F. Percent Composition

The percent composition program can remove tedious calculation. It is also a program that is not needed as a teaching tool. Programs [13, 23] calculating the percent composition of a sample or the empirical formula of a compound once the laboratory work has been completed, simply utilize the computer as a big, expensive calculator.

## G. Computerized Homework

Several instructors have written programs for computerizing homework assignments [12]. Because of the tremendous number of man-hours devoted to correcting homework, this is one of the first areas tried by the novice CAI user. Very limited success has been achieved in this area. Many students resent having their work corrected by an impersonal computer and also consider the making of computer cards an added assignment. Not only do the poor students fail to turn in the work, but many of the average and slightly above average students also. Thus, the benefits of computerized homework seems to be limited to the top 10 to 20 percent of the brighter students. Instructors often argue that it gives the students a chance to submit problems twice; this is true, but the students do not usually avail themselves of this option.

Success is achieved when the students can receive tutorial help from the computer [24], but this type of CAI homework demands a tremendous number of man-hours in preparing programs by the instructors and requires a sophisticated computer system (e.g., PLATO), which is not normally available to the small school.

Repeatable exams [25] have been very successfully used at the University of Indiana, but this is due partly to the fact that the instructor does all the work in preparing the tests and the students hardly know that the work is computer-generated. In like manner, the computer has been successfully used to grade unknowns [26]. Chappell and Miller gave each student a deck of cards containing his desk number and cations found in the qualitative analysis scheme. At the end of the laboratory period the student returned those cards which corresponded to the ions in his unknown. These cards were then graded on the computer. The grades were posted and cards returned to the student for future use. Although the documentation on this idea does not explain why the program is successful, it is probably due to the elimination of the student's need to go to the computer center or keypunch to report his data.

## III. SIMULATION PROGRAMS

The greatest use of simulation-type programs has been in the smaller colleges and universities. This is due in part to the lack of readily available instruments, or in larger schools to counteract empire building when a faculty member feels that only his students are competent enough to use his equipment. It is evident that many of the simulation programs were first developed by the researcher in attempting to interpret his data. While a large number of programs are available to the educator from the researcher, the transfer from research to education use is often more theory than practice. The researcher is familiar with the instrument and knows what input is needed and what kind of output is expected; this is not true with students. The programs must be written so that the students can learn as much from a simulation situation as from the experiment in the laboratory.

Although a large number of simulation programs exist in the literature or through distribution centers, the documentation is poor. The number of man-hours required to transport some of these programs virtually prohibits their use; unlike number-crunching programs, where only the answer is important, versatility is paramount.

### A. NMR

Almost everyone involved in simulation programs eventually writes an NMR program or procures one from someone else. A large variety of programs are available [27] and some of these are extremely well documented.

NMRs are expensive to acquire and just as expensive to run. Thus, the simulation programs allow the students to obtain spectra which show coupling constants, splits, shifts, etc., and give them a little "on-line" experience. In a canned program, there is the limitation of data input, but unlike taking the data from the literature, the students get to handle spectra and can try to identify compounds just as if the data were run on an actual NMR.

### B. Mass Spectrometer

Most other chemical instruments have been simulated on the computer. A mass spectrometer program was published in 1968 [28] in FORTRAN II and was revised and prepared in the more transportable FORTRAN IV in 1971 [13]. A simulated mass spectrometer allows the undergraduate students to perform operations which they never would be allowed to perform in the laboratory, because of probable instrument damage. The simulated instrument allows the instructor to use any combination of isotopes he desires, most of which would be too expensive to use in the laboratory. The simulated instrument allows the students to play with the instrument, and the playing can be very real if the students can sit at a teletype or computer console and vary slit width and voltage until he receives what looks like a real spectrum.

The simulated mass spectrometer program [13] is provided with real information, including random noise; thus, the students learn to distinguish the difference between an isotope and the noise of an instrument. Once the students have obtained the spectra, they still need to identify the isotopes and do the same calculation required in the laboratory. Thus, the learning process is twofold: (1) learning to use the instrument, and (2) performing meaningful calculations on the data. If the calculations are performed for the students once they have obtained the spectra, they rarely remember what they have learned, and the simulated program is not any more useful than reading a textbook.

A modification of this idea is available for organic metallic compounds [29].

## C. Identification Games

### 1. Organic

Another type of laboratory simulation program is available--the identification of laboratory data in the game form. A great deal of reaction chemistry and systematic analyses can be taught with these games, and if the competitive nature is emphasized, the students put forth more effort.

A unique type of organic identification game has been developed for the PLATO system [30]. However, the sophisticated PLATO is available to only a few individuals and an organic identification game has been written in card deck form [31]. This game has 42 organic qualitative analysis tests and data information sets (41 tests and the services of a consultant). The consultant service is unique, and should help the solution of an unknown and create "conversation" in the use of the game; however, documentation on the consultant service is unavailable and its application to the overall game seems to be untried. Some of the tests available are m.p., elemental analysis, solubilities, common derivatives, mass spect., UV, IR, and NMR. To preserve the competitive nature of the game each test is given a dollar value, estimated to be the amount of money charged by a trained analyst in obtaining the results in a well-equipped laboratory. Although every good organic and analytical chemist knows that a pure compound can be identified by mass spectrometry, IR, and NMR, the cost of these tests is high;

the unknown compound can be identified for less money by solubility tests and elemental analysis. Part of the game is to teach students to use discretion in requesting tests. An attempt is made to teach a student that two or three simple tests from an analyst may give him the same results as the more expensive instrumental tests.

The game has been beaten by students [12]. They learned that an IR is a relatively inexpensive test. Since they were not paying for computer time, they called for the IR bands, retyped the data, and ran it on the ISIS program described earlier. In many cases, ISIS returned one compound, the correct one. In the cases where ISIS produced more than one compound, the correct one was identified with an elemental analysis. After this technique was learned by the students, the game lost its usefulness.

The disadvantages of the organic game [31] is the large amount of CPU time it requires, and because the game stores all its data in the computer core, the core requirement prohibits using the game on a small computer.

2. Inorganic

Other laboratory games are available in the inorganic qualitative analysis area [32, 33]. These games are primarily written to help teach qualitative analyses rather to provide a "fun" experience. One of the inorganic games [32] is structured to use 67 cation tests to identify the normal cations taught in a freshman laboratory; another [33] is an unstructured game, avoiding redox equations, to teach the identification of fewer cations.

The structured game, written in FORTRAN IV, provides the students with the same shortcuts allowed in the laboratory and penalizes them for bad judgment. It has been designed to allow changes by the individual instructor. It can be segmented to run with 8K of core. The biggest disadvantage in both of these games is that documentation on the students' reception is missing.

## IV. CONCLUSIONS

Although there are many canned programs available from the decomposition of $KClO_3$ [13] to working of the First Law of Thermodynamics [13], the whole approach to using canned programs lies with the individual instructor. It provides an avenue for him to teach more and emphasize deeper concepts of principles. It provides the instructor having only a meager knowledge of computer programming and operation the opportunity to use the expertise of the skilled programmer to enhance his curriculum without spending very much in background preparation.

The biggest problem in the area of CAI with canned programs is obtaining information. Until media of exchange are provided, the complete use of CAI will not be available to the interested instructor. Lack of transportability also needs to be overcome, but this problem will not be so formidable if better methods of documentation are provided.

## REFERENCES

1. H. Cassidy, J. Chem. Ed., 49, 34 (1972).
2. J. A. Young, J. Chem. Ed., 47, 758 (1970).
3. Proceedings of Conference for Computers in the Undergraduate Curriculum, June 16, 17, and 18, 1970, University of Iowa, Iowa City, Iowa. (Iowa Conference.) Proceedings of Conference for Computers in the Undergraduate Curriculum, June 24, 25, and 26, 1971, Dartmouth University, Hanover, New Hampshire (Dartmouth Conference.) Preliminary Proceedings of a Conference on Computers in Chemical Education and Research, July, 19, 20, 21, 22, and 23, 1971, Northern Illinois University, DeKalb, Illinois. (N.I.U. Conference.)
4. P. C. Jurs, B. R. Kowalski, and T. L. Isenhour, Anal. Chem., 41, 21 (1969).
5. J. R. Denk, N. I. U. Conference [3], as reported in C. and E. N., Aug. 2, 1971, p. 7.
6. H. Shull, N. I. U. Conference [3], p. 7-2.
7. L. J. Soltzberg, J. Chem. Ed., 48, 449 (1971).
8. E. D. Smith and D. M. Mathews, J. Chem. Ed., 44, 757 (1967).
9. R. D. Nelson, Jr., M. R. Ellenberger, and F. R. Ellenberger, N. I. U. Conference [3], p. 1-21.
10. J. R. Denk, N.C.E.C.S., Research Triangle Park, North Carolina, private communication (1970).
11. J. E. House, Jr. and R. C. Reiter, J. Chem. Ed., 45, 679 (1968).
12. L. R. Sherman, N. I. U. Conference [3], p. 6-13.
13. Program and Literature Service (PALS), N.C.E.C.S. Research Triangle Park, North Carolina (J. R. Denk, ed.), August 15, 1971.
14. J. W. Drenan, J. Chem. Ed., 32, 36 (1965); E. B. Thomas, ibid., 40, 70 (1963); A. R. Emery, ibid., 43, 131 (1965); G. G. Schlessinger, ibid., 46, 680 (1969).
15. M. Bader, N. I. U. Conference [3], p. 9-14.
16. K. J. Johnson, N. I. U. Conference [3], p. 6-5.
17. P. Blatz, D. Pippert, L. R. Sherman, and V. Balasubramanigan, J. Chem. Ed., 46, 512 (1969).
18. J. R. Platt, J. Chem. Phys., 25, 80 (1956).
19. Written by K. J. Johnson, University of Pittsburgh, Pittsburgh, Pennsylvania.

20. J. R. Denk, N. I. U. Conference [3], p. 10-76.
21. K. J. Johnson, PALS [13]; J. Watson, PALS [13]; W. Gasser and J. L. Emmons, J. Chem. Ed., 47, 137 (1970); G. F. Luterie and J. M. Denham, ibid., 48, 670 (1970).
22. H. Hermanson, North Carolina Agricultural and Technical State University, Greensboro, North Carolina, unpublished (1971).
23. Written by L. R. Sherman, North Carolina Agricultural and Technical State University, Greensboro, North Carolina.
24. R. C. Grandey, J. Chem. Ed., 48, 791 (1971).
25. F. Prosser and J. W. Moore, N. I. U. Conference [3], p. 9-26.
26. G. A. Chappell and R. M. Miller, J. Chem. Ed., 44, 79 (1967); K. M. Wellman, ibid., 47, 143 (1970); S. G. Smith, R. Schor, and P. C. Donahue, ibid., 42, 224 (1965).
27. M. Bader, J. Chem. Ed., 48, 175 (1971); C. L. Wilkins and C. E. Kloppenstein, ibid., 43, 10 (1966).
28. T. R. Harbros and C. W. Miller, Computer-Based Physics, Commission on College Physics, 4321 Hattwick Road, College Park, Maryland (1969).
29. B. D. Domlek, J. Louther, and E. Carberry, J. Chem. Ed., 48, 729 (1971).
30. S. Smith, N. I. U. Conference [3], p. 4-39.
31. F. Hornack, Dartmouth Conference [3], p. 359.
32. L. R. Sherman, The Faculty Review, Bulletin of the North Carolina Agricultural and Technical State University, Greensboro, North Carolina, 65, 52 (1973).
33. G. A. Gerhold, West Washington State College, Bellingham, Washington, private communication (1971).

Chapter 8

# COMPUTERIZED HOMEWORK PREPARATION AND GRADING

N. Doyal Yaney

Department of Chemistry
Calumet Campus Section
Purdue University
Hammond, Indiana

## I. INTRODUCTION

For the sake of simplicity, I will assume that the reader has no previous experience with data processing, other than that which he may have obtained from his telephone bill, gasoline credit card account, or a U. S. government check, in the form of a punched card bearing the admonition "Do not fold, spindle or mutilate." This is because most of my contemporary scientists received their formal educations before the computer age, or received it from teachers who received theirs before the computer age, and as a result now have little knowledge of data processing and often shy away from it.

While I have empathy for my humanities friends and their interminable jokes about computer mistakes, I would exhort them to consider the advantageous possibilities of this new form of communication--especially of its saving in repetitive record-keeping drudgery even for their own disciplines. As an example, a form of my "Custom Homework" can take on the aspects of a one-dimensional crossword puzzle. To my fellow educators, consider your students, and the fact that they will have to spend their working careers living in, or working with, this computer age.

As humans, we all have our fears of the unknown. We should not allow education to be the last reactionary bastion to change; instead, we should strive bravely to learn the fundamentals of the new unknown and expose our students to it at the same time to learn along with us.

There is an abundance of esoteric literature at the sophisticated level of postgraduate and graduate research, and even occasionally in upper level undergraduate curricula. There is a paucity of it at the lower undergraduate level, although high schools, junior highs, and even grade schools are beginning to expose their students to computers and data processing.

If we are to meet the challenges to education in the 70s, we will need all the efficient tools we can get our hands on, so that we can devote more of our energies and time for creatively meeting the needs of our students more effectively. Herman Hollerith recognized the challenge of a decennial census-taking requiring longer than ten years to process back before the turn of the century, and devised the code we use on punched cards to provide input to our computers.

We all recognize that educational time is stretching longer and retirement sooner, so that the twain shall meet. More must be acquired on an informal do-it-yourself basis, with quick access to expert tutorial advice from machines of all types or personally with teachers when needed.

Thus, I shall begin by considering the ubiquitous IBM card and its punch codes, referring generously to publications by IBM and others for additional details you may wish to consult on your own to supplement what is described here.

If you are familiar with the standard typewriter keyboard, you will adapt readily to a keypunch machine. A useful Reference Manual is IBM's A24-0520-2 for keypunch Models 24 and 26 [1].

Keep in mind that all data processing equipment expects to find similar type data in the same fields of card columns for all cards involved in a particular job application.

It is almost axiomatic that for most applications of computers that entire systems of data processing must be instituted in toto to function effectively. There are trying times during the debugging process that inevitably results. Also, unless your department has a budget large enough for its own computer and associated equipment, most operations will have to be done on a shared basis with others at a central computer installation. This works reasonably well for most of the peripheral equipment, but often it is difficult to obtain on-line (walk-in) service on the central processing unit (CPU) of the computer during prime time; its schedule is limited, etc., etc., as the King of Siam was prone to say. Having been through these problems during the development of this system, and for that matter still struggling with these and other red-tape problems, I can sympathize with those who have been wary of joining the computer age during its growing pains as novices, especially if they have previously had unfortunate experiences in this regard.

The system was developed stepwise, retaining the option of working at any stage either manually or by machine with minimal loss of effectiveness, and, where equipment permitted, leaving alternate paths of processing available--especially allowing the option of utilizing associated data processing equipment when the computer was not readily available, for whatever reason.

Using punched cards as the primary initial means of providing input records for the machines is one of the best ways of having your cake and eating it too--in fact, one might say two. This is because cards have a dual purpose--they can be easily read both by humans and by machines. In this way one can switch readily from manual to mechanical operations and vice versa, as the situation dictates. In addition, cards are not easily wiped out, demagnetized, etc., and can also function as your permanent hard-copy file.

The purpose of this chapter is to be useful directly to noncomputer-oriented instructors, and accordingly I intend to err on the side of too much explanation rather than too little, i.e., generalization. While unit record and computer manuals may be consulted for some of the information included here, I intend to make this work stand on its own, permitting the readers to prepare their own systems or portions of a system at their own pace, with a minimum of searching in user manuals for the details required for their particular use of applications.

For those readers familiar with data processing equipment, Sections II and III may be largely redundant, but for the uninitiated it is hoped that the descriptions and pictures will help them to view their respective computer facilities with a new perspective and appreciation, rather than with awe and bewilderment. They may find that they can even carry on a conversational dialogue with data processing personnel without going away mumbling in their beards following what usually turns out to be essentially a monologue.

The organization of the chapter is essentially chronological. Communication with most machines is commonly via keyboard. The keyboard may have direct access to the computer, while simultaneously producing printed copy and punched cards or paper tape. Alternatively, the keyboard may be used as a stand-alone device, as typified by a keypunch or even a teletype machine.

The punched card is the main transferring medium between the keyboard and the equipment that we will discuss, primarily because it is in rather widespread use. Corrections, deletions, additions, and rearranging are easily accomplished with cards, in contrast to paper or magnetic tape. The discussion then proceeds logically to equipment requirements (Section III), including some of the optional possibilities. With these preliminaries accomplished, the readers are introduced to the heart of the system, the development of the library of questions and answers (Section IV) which is the source material for generating the "Custom Homework" (Section V). (This leads to the need for solution of any calculation problems associated with the homework--computer programs for student calculations are available from the author.) There follows a discussion of machine grading (Section VI), and lastly, subtotalling and final totalling of grades (Section VII).

While the organization of this chapter proceeds from the initial input of questions and problems via the punched card to the final grade summation, the various sections may be used as individual units or arranged in various combinations to produce logically operating subsystems.

Our campus has a standard computer exam-grading program for handling the usual multiple-choice-type questions (maximum of five choices) for which all the choices had to be printed as a part of the the text of the question itself. Traditionally, this type of question encourages guessing rather than the more pedagogically desirable logical reasoning or deduction, especially in the realm of calculable problems requiring arithmetic answers. Thus, the system described here requires the student to generate his answers rather than choose from five preselected alternatives. To more fully utilize the punched card output of grades from the standard grading program, I chose to program the accounting machine (IBM 407) by wiring a board to read the grade cards, list them, and sum the scores for each student and for the class. Subsequently, programs were written for the computers available. Thus, I integrated grade totalling (as described in Section VIII) with the punched card grade output from the predecessor of the grading programs described in Section VII.

Coincidentally, I set about taking advantage of the simple punched card (Section II) to alleviate the logistics of handling and record-keeping of quiz answers and their scores for hundreds of students. Initially this involved punching student number, name, course number, and section on cards distributed for answering subjective questions to be graded manually or objective questions for machine-grading. The cards are easily duplicated for subsequent quizzes, readily machine-alphabetized for quick distribution to large classes, both before answering and after grading. They are excellent source material for quickly printing up-to-date rosters, as well as rosters of scores from quizzes and exams.

It was the disadvantages of the multiple-choice grading program which led to the development of the more versatile system described in this chapter. Chief among these disadvantages is the lack of variation in credit for questions of widely varying difficulty. Each answer was fixed at one point (or a fixed percentage of the total) regardless of whether it was a simple true or false or a complicated chemical mathematics problem. In this system the point credit is equal to the length of the answer, that is, to the number of digits or characters in the answer. A single character answer, such as + or - is worth one point, a four-character answer four points, and a nine character answer nine points--all tailored to the desired weighting.

## II. THE PUNCHED CARD--ITS FORMAT AND USES

### A. The General Purpose Punched Card

One type of card is shown in Fig. 1 (IBM 5050). It consists of 80 vertical columns, each accomodating a digit or character, such as a letter of the alphabet. Thus, each card can carry a short sentence or line of print. Machine-readable informational holes or punches may occur in any of the 80 columns at any one of 12 different levels or horizontal rows. For numbers, the digits are represented by punches in one of the ten levels from zero through nine, the latter at the bottom of the card. The twelfth and eleventh are at the top and next to the top, respectively, of the card immediately above the zero (or tenth) level or row. These three rows, twelfth, eleventh, and tenth (or zero), collectively known as a zone, arc punched to selectively represent the first, second, and last third of the alphabetic characters A through Z in conjunction with one of the nine digit punches. Thus, alphabetic characters are represented by double punches per column, in contrast to single punches per column for digits of numbers. As you look at the card from left to right you will note a systematic pattern related to the preprinted information on the card. Across the top of the card above the 12-punch row will appear the electric typewriter printed character represented by the punch or punches appearing directly below it. In our Fig. 1 alternate columns are used to improve readability, beginning with a 12 punch (+) in card column (cc) two, followed by an 11 punch (-), the digits zero through nine, and then by the three portions of the alphabet, A though I, J through R, and S through Z, with the slash, /, in card column 62.

The first convenient use found for the punched card is to prepare a master deck containing the names of the students of each class. If your school is set up for it, these may be obtained from computerized enrollment files. If not, they may be readily keypunched and used repeatedly to produce rosters for class seating, lab unknowns, to record grades, to make additional duplicate decks, etc. The master card might include additional information of use to the instructor not necessarily keypunched into the card. Such helpful information might include: where and when previous courses in the subject had been taken, outside employment hours per week, major field of interest, emergency phone numbers eyesight or hearing difficulty or medical disabilities which could be critical in a laboratory emergency.

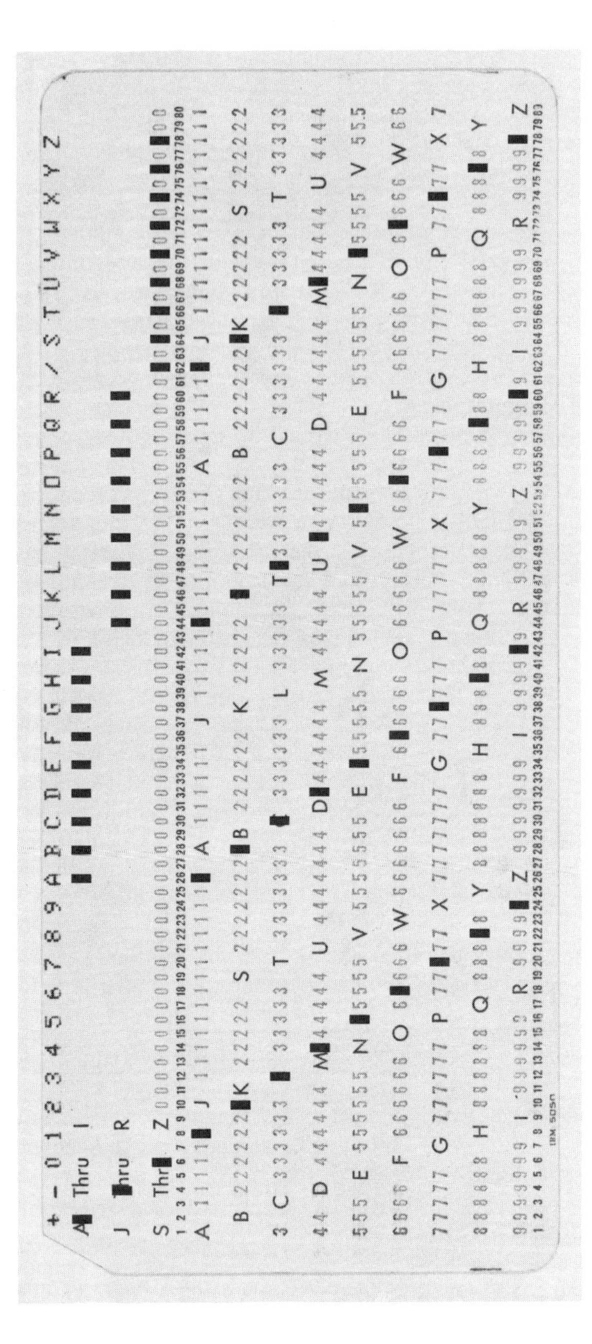


Fig. 1. The general purpose punched card.

The master card might also be utilized for recording an entire semester's grades if one does not wish to avail oneself of other alternatives suggested in the following sections. These grades could then be machine-totalled by computer or IBM 407 (see Section VIII), or manually totalled from an easily prepared printout.

Figure 2 shows a typical master card for a student. We use a six-digit number assigned alphabetically [2] (see IBM Accounting 10,000 Division Code for Proper Names, X-21-5114) as a student number in cc one thru six, the last name beginning in cc seven, or optionally in cc eight for easier reading, followed by initials or preferred nickname to personalize the individual. The division or section of the course is punched in cc 24 and 25, and the course designation and/or number in cc 26 through 29. The remaining columns may be used for a variety of purposes on various duplicate decks.

A very convenient use for quizzing large classes is to have a deck duplicated in advance, and set out in several piles (ten according to first digit, or 26 by alphabetic order) at the front desk for immediate self-serve pickup by each student. Any type of question may be given if the answers are to be written on the back of the card for manual grading later, or if machine grading is desired, mark sense or Port-a-punch cards may be used. The latter two types may be hand graded by using overlay key cards punched with larger holes for the mark sense answer cards or with regular IBM punches for the Port-a-punch cards. The latter two types of cards might be used several times for short quizzes that fit totally within the capacity of the answer card. Alternatively, they may be hand-graded and the grade punched into unused card columns for subsequent machine grade recording, posting, and tallying. The cards are easily machine-alphabetized again for ease in returning to the students for desired checking, and ultimately become part of your machine-readable record file for the class.

A corollary advantage of utilizing this system is having an attendance check each time a quiz is given or the grade card is returned for checking.

### B. Special Purpose Cards

These cards are identical to the general purpose cards as far as the machines which read them are concerned, but they have specialized adaptations for the user.

#### 1. Mark Sense Cards

These cards are designed so that marks made by soft graphite pencils may be read by machine. The placement of the mark determines the character just as the position of a punched hole does in a card. Most commonly,

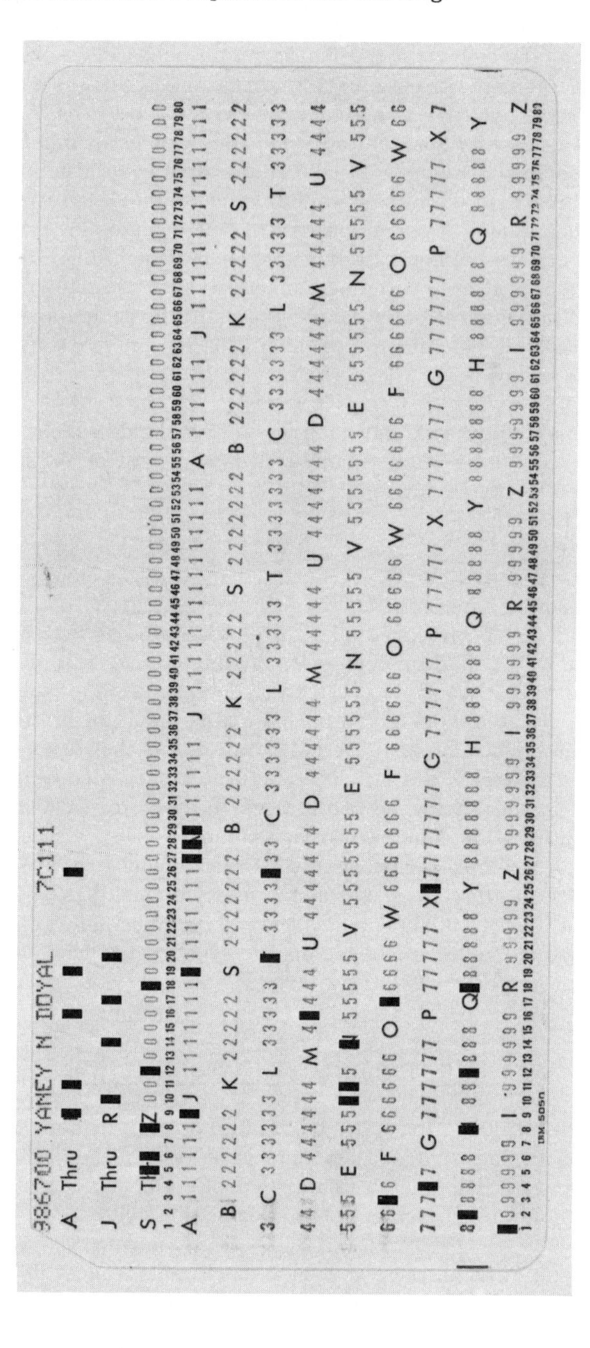

Fig. 2. Typical student master card.

machine readers convert these marks into punches in the same card, but some are designed to transmit the information directly to a computer. Figure 3 shows one type of mark sense card which is readily available (IBM U66887); it is readily adaptable to recording grades or other data as previously described. Figures 4 and 5 illustrate a card specially prepared by the author to utilize both sides so that twice as much information is encoded. Normally one side of a mark sense card accomodates 27 characters, since each mark covers the equivalent of three punch positions. Both sides can accomodate 54 characters; some computers split the information horizontally on the card and theoretically handle 108 multiple-choice answers per card, while still leaving 26 columns for identification information such as student number and name.

Marks may be erased, but no stray marks should appear anywhere on the face of the card to be read. Signatures may appear on the reverse side if it is NOT used for mark sensing.

### 2. Port-a-Punch Cards

These are the same size as general purpose cards, but half the card columns, the even-numbered ones, are perforated so that the punches may be removed by a pencil or pen point, making the cards immediately processable without the necessity for mark sense conversion to punches. I caution students to circle the desired answers and recheck them after completing the card before punching out the perforations, as they are not so easily erased as mark sense cards. The perforable card's capacity is 40 characters. The 40 odd-numbered columns can be used for student number, name, course identification, homework page numbers, etc., as desired. Figure 6 shows one of the more useful forms of Port-a-punch card (IBM D10687). Figure 7 is a similar card, but lacks the field markings which aid the student in keeping track of his punching. It would be more desirable to have the Port-a-punch columns numbered consecutively 1 through 40 rather than using the even card column numbers 2 through 80.

## III. EQUIPMENT REQUIREMENTS

As indicated in Section I, time availability on your computer may be limited, so it is advantageous to know what part of the processing can be done on separate machines which are specifically job-oriented, known in the trade as unit record data processing equipment.

Some form of an electric keyboard machine is almost mandatory, either directly accessed to the computer, or as a separate device such as a keypunch or teletype. Since I used a punched card system with the library of

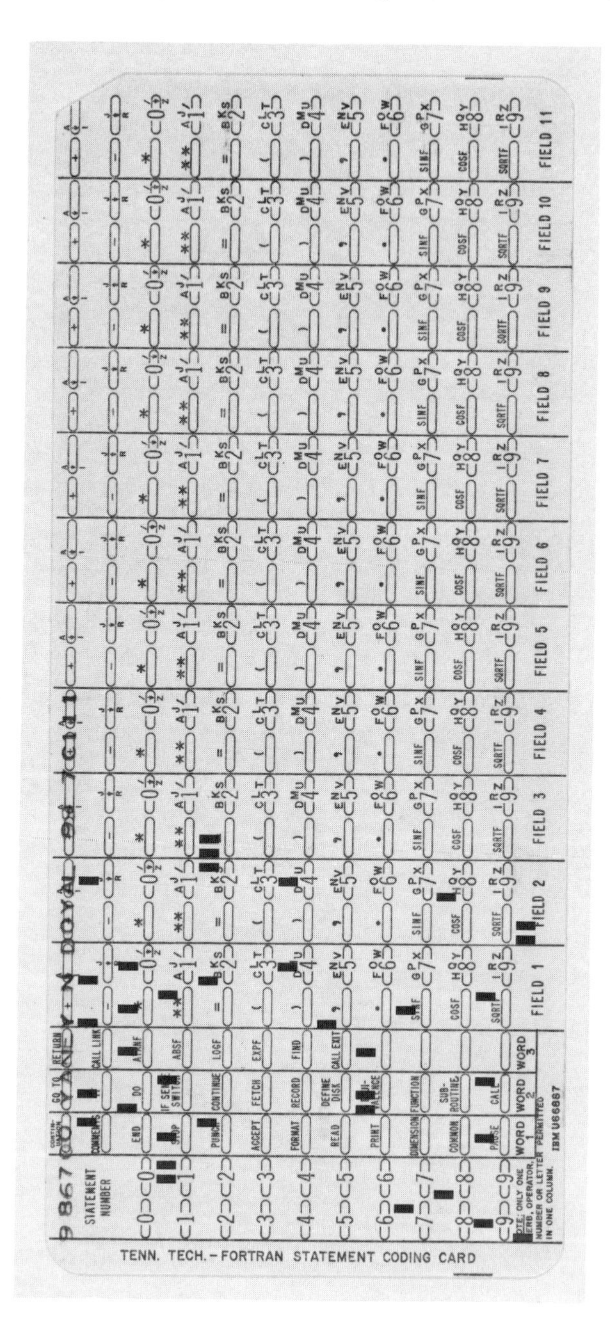


Fig. 3. Mark sense card, general purpose.

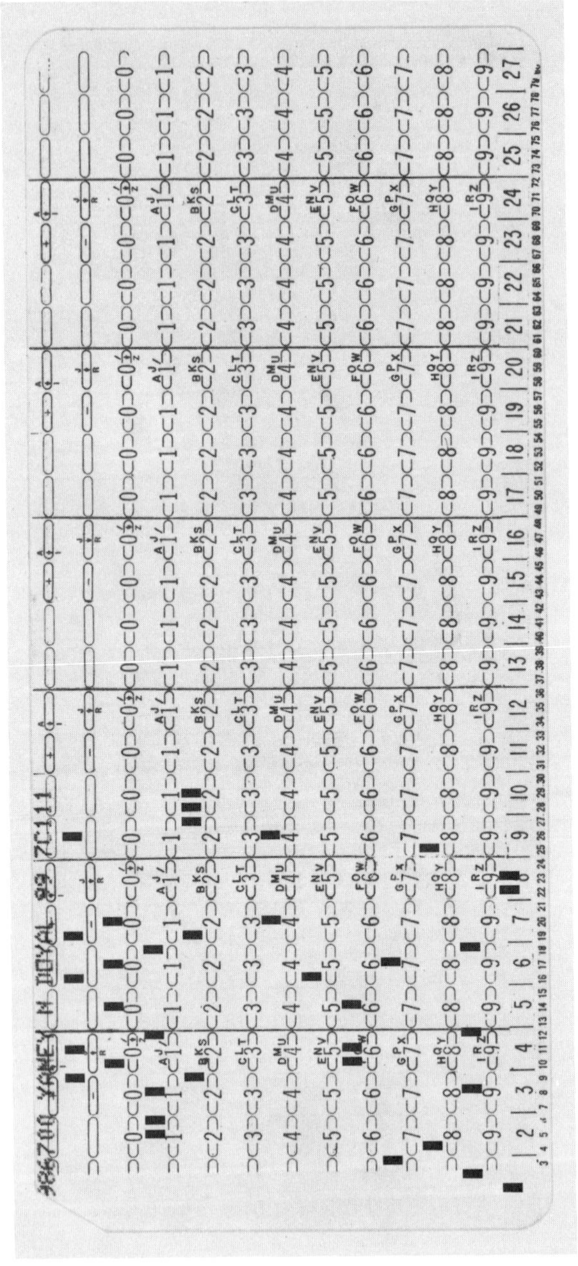

Fig. 4. Mark sense card, special purpose, front side, 27 positions.

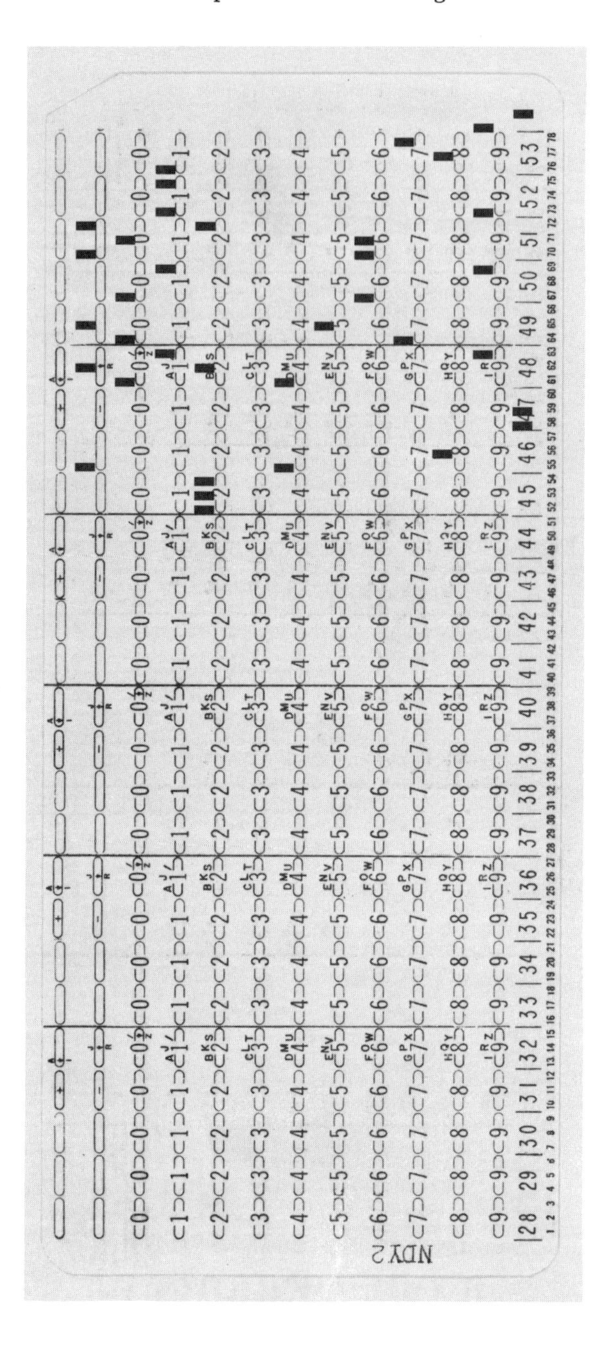

Fig. 5. Mark sense card, special purpose, back side, 54 positions.

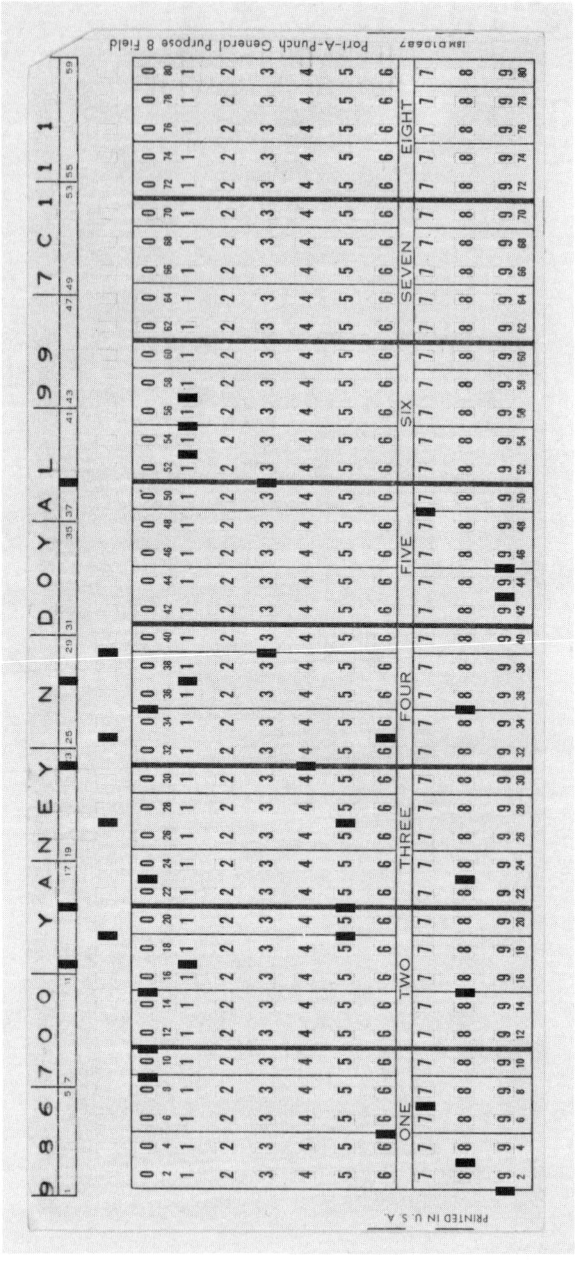

Fig. 6. Port-a-punch card, special purpose, eight fields.

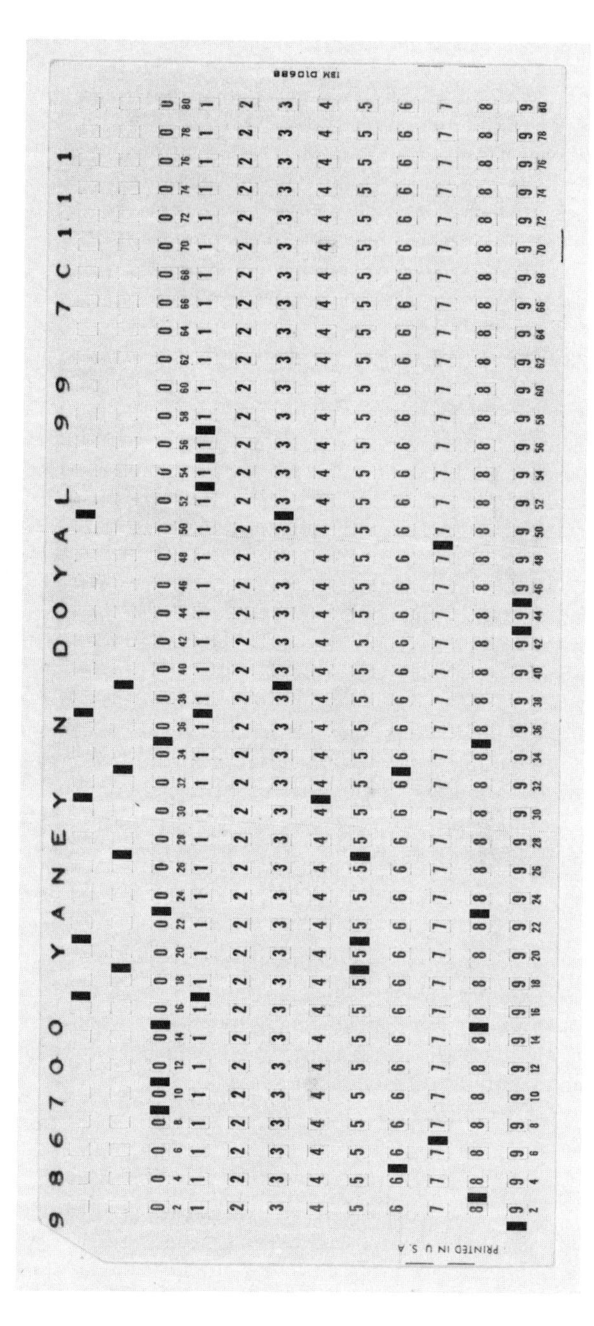


Fig. 7. Port-a-punch card, general purpose.

questions directly under my own personal security, this discussion will obviously be oriented in this direction. But the questions can just as easily be placed on punched paper tape, which could be stored on magnetic tape, disks, drums, or any of the similar sophisticated computer memory storage devices. This poses somewhat different problems of security, especially if students are hired by the computer center as operators. If these methods are employed, sorting for questions in a particular category may be done internally by the computer and the appropriate questions printed as desired. With magnetic disks or other random access storage devices, it is of course possible to set up random number generators which could select questions at random and which could be further checked as fitting the proper category or categories requested, prior to and as a condition for printing as homework.

While almost everyone prefers working with the latest and most sophisticated computers, it is often more convenient to print a roster of manually graded scores at 150 cards (names) per minute on the 407 accounting maching than wait for the computer to be available.

Computer manufacturers concentrate primarily on improving their more sophisticated equipment, especially so it can be more fully utilized as its internal speeds are maximized. This requires multiple input devices, involving many remote terminals, such as are utilized in chain department stores or airline reservation offices, or peripheral computers and their associated disk, tape, or card readers. If you are fortunate in having this type of equipment at your installation, the library of questions may be accessed remotely by the students with the computer handling all the record-keeping for you. However, if you are limited in the facilities available to you, this system is intended to demonstrate what can be done with minimal computer facilities and commonly available unit record data processing equipment.

Regardless of the type of facilities available, the questions need to be designated so that suitable answers may be recorded by mark sensing, Port-a-punching, regular keypunching or other keyboard response. These questions and their answers need to be keypunched or otherwise electrically communicated to the necessary equipment.

## A. Punches

1. A hand punch for really portable use is a heavy duty paper punch modified with an IBM-sized die. This punch has a two-inch reach which allows access to any columns and rows of the card. It is pocket-sized, and handy for punching grades into manually graded quiz cards, etc. It is available from McGill Metal Products Co., Marengo, Illinois for about ten dollars.

2. A manual punch with a card column indicator and spring loaded advancing mechanism is available in a 12-key nonprinting model for a few hundred dollars from the Wright Line Co., Chicago, Illinois. An alphanumeric printing model using a dial selector system similar to a Dymo tape writer sells slightly higher.

3. IBM has manufactured an electrified, nonprinting 12-key Model 10.
4. The most suitable keypunches are IBM Model 24 nonprinting, or (preferably) printing Model 26 or 29, all with numeric, alphabetic, and special character keys. These are the most widely used types of electric keypunch machines. They cost a few thousand dollars each, or they may be leased for a monthly rental charge. The Model 29 may be obtained with an interpreter feature which reads an unprinted card and prints directly upon it. This is important for printing on duplicate decks unless you have a separate interpreter, since most computers and reproducers do not print as they punch cards. Cards reproduced directly on these keypunches are normally printed simultaneously with the punching, except for the non-printing model.

## B. Sorter

A mechanical sorter of the IBM Model 82 or similar type sorts one column at a time into one of 12 receiving pockets representing the 12 horizontal rows of the card. In this way the units, tens, hundreds, thousands, etc., positions of a new number field can be systematically sorted so that a deck is then arranged in ascending sequential order. If the numbers are assigned in alphabetical order, then alphabetization thus results. Alphabetical sorting can be done on last names all starting in the same card column, but double sorting must be done on each comumn, one time on the digits 1 through 9, and then again on the zone punches only, by setting the sorter accordingly.

As indicated earlier, the sorter is very convenient for putting quiz, test answer, or roster cards into sequential or alphabetical order. I have also found it useful for arranging grade scores into sequential order for ease in determining letter grade division points and for determining quickly the median score by dividing a deck into two parts of equal thickness, or by printing in roster form and folding the long list in half. No wiring boards are used for the sorter operation.

## C. Collator

A collator of the IBM Model 85 type is useful for merging two decks of grade cards into sequential order. It is especially useful if each deck is sequentially ordered, as several columns are checked simultaneously rather than a single column at a time as in the sorter operation. In addition, it checks for errors in sequence, which the sorter does not do.

I find it useful for combining several decks from different homework, quiz, and test score cards which have been previously sequenced for computer grading and printouts. Aside from sequence-checking and some other features, the collator can be considered as optional equipment if a sorter is available. The simple wiring required for the collator is described in [3] IBM Reference Manual A24-1005-2, and an example is shown in Sec. VI.A2.

## D. Interpreter

An interpreter reads the holes punched within a card and prints it upon the same card. It is used on decks duplicated by computer or by the reproducer described below, as most of these devices do not incorporate the printing function. The print is larger than that obtained on a keypunch, making it more easily read, and therefore only 60 characters can be printed across the card in any one pass. The remaining 20 columns require a second pass using another wiring board and need to be printed on another line of the card. Obviously these larger-size characters cannot be aligned directly above their respective punches in the card columns. Most installations will have boards already wired to print all the first 60 card columns on the print line you select, and the last 20 card columns on another print line which you must remember to change when you slip the other board into position. The simple wiring required for any other selected printing you might conceivably desire can be found in IBM Reference Manual for Models 548 and 552 and numbered 224-6384-2 [4]. The Model 548 has a 60 card/min. speed and prints only on rows one and three near the top of the card. Model 552 has a 100 card/min. speed and prints on any line selected from one through 25, the odd lines being between the punches, and the even lines directly on the punch rows. An interpreter can be considered optional if your computer has an interpretive feature incorporated into its card reader, or if your keypunch has a similar feature.

## E. Reproducer

A reproducer of the IBM Model 514 type will punch a duplicate deck if you feed the original deck into the left read hopper, and a deck of blank cards into the right punch hopper. All 80 columns can be read and punched in one pass through the machine. Again most installations will have what is called an 80-80 board to accomplish this. It is usually wired to simultaneously check the punching against what has been read and indicates any card columns not matching. Special wiring can be done readily to read selected columns and to punch these at any column location in the receiving card. I commonly use this feature to transmit the student identification from cc 1 through 29 in master cards or homework keycards to the odd-numbered card columns 1 through 57 in Port-a-punch cards for homework, lab reports, and tests (see Section V.B.1.d).

An optional feature available for the reproducer is necessary for converting the special graphite pencil marks made by students on mark sense cards into punches in the same cards in columns determined by a wiring board (see Section VI.A.1.a, or IBM Reference Manual A24-1002-2)[5]. This feature is available usually in multiples of nine positions, 27 being required to read the entire face of a card. If only nine positions are installed, three passes of the cards using an altered wiring board on each pass would be required to read all 27 mark sense positions on one side of a card. If both sides of a card are marked, a minimum of two passes per card are required.

A reproducer is not required if all duplication is done by a keypunch or computer and if no mark sense cards are utilized. They can also be read by optional computer input readers (such as IBM offers), reading 40 marks per card (formatted similarly to the perforations on a Port-a-punch card), thereby increasing the capacity and more importantly requiring only one pass to read and process both punched and marked information.

## F. Accounting Machine

An accounting machine of the IBM 407 type is, for our purposes, a convenient printer not requiring an associated computer to control it. It is a self-contained card reader and page printer with the ability to recognize different punched card types and process them accordingly, perform arithmetic functions, store limited amounts of information to be printed on demand (such as headings for pages), and to emit impulses to print characters not contained on the input cards.

It can be used to sum 11 scores, all punched in a single student master card, and total them (called crossfooting), printing all information from the card as well as the total (see pp 150-153 of IBM Reference Manual A24-1011-2) [6].

It can add scores punched singly per student card, print all desired information from each card followed by a subtotal for each student, count the cards or the students (i.e., number of subtotals), and at the end print the number of cards, student subtotals, and the final total of scores.

For the homework application, the IBM 407 is used to print the questions, space the answers on the printout for detachment as the instructor's key, number each question incremented by its point value, and consecutively number each page both on the question and on the answer portion. Both portions of each page are also dated from information stored in the machine from the first card.

My date card carries the two digit day in cc 17 and 18, the month in cc 19, and the year in cc 20. I use the digits 1 through 9 for the first nine months designation, zero for October, N for November, and D for December, thus requiring only one column for the month. This and the year field may be expanded if desired by adding a few wires to the wiring board or by modifying the computer program for the date processing (Fig. 8). The date card is recognized by an 11-zone punch (-) in cc 14; accordingly, no other question cards may have such a punch present, although numeric punches may be present for classifying and identifying the question in this card column.

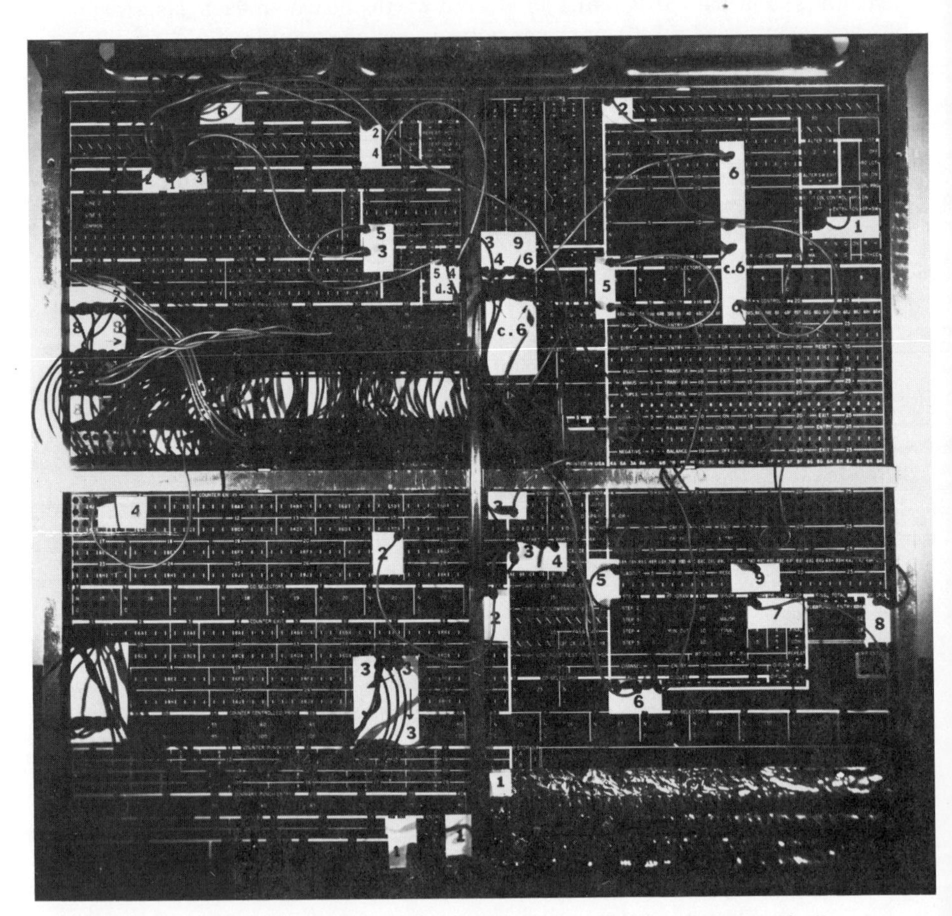

Fig. 8. Accounting machine, wiring board for "custom homework."

Especially valuable is the capability of easily printing Ditto masters on the IBM 407 by selecting the question cards desired for your tests, and passing them through for processing in a few minutes.

### G. Computer

Of all the equipment described, the computer is the most versatile, the most trouble-free, and the most utilized--all of which means that any installation will present its problems regarding access time. It reminds me somewhat of the family entertainment center which encompasses into one unit FM, AM, TV, record player, tape deck, etc. It becomes difficult to utilize several functions simultaneously, and thus it is with many computers--especially those lacking multiple input and output devices and time-sharing central control. While our Multi-Function Card Machine (MFCM Model 2520) can read, punch, sort, collate, merge, sequence, interpret, and cause the attached printer to print under the computer's control, it cannot do any of these for me while performing another job. Finally, like a breakdown in the entertainment center, a failure in the computer or even in one of the devices may make the entire system inoperative and leave you high and dry. Therefore, I have found it advantageous to have many of the simpler functions conducted on data processing equipment not assigned (dedicated) to the computer's control.

## IV. LIBRARY OF QUESTIONS AND ANSWERS

This section gives specific examples of the types of questions and especially of the type of answers best suited for objective questioning. For the most part these questions are oriented toward numerical responses, although combinations of numbers and letters are utilized in asking for chemical formulas as answers. Some complete, alphabetical, fill-in-the-blank type word answers are included to demonstrate their feasibility, although it is realized that this requires double punch-coding per card column. Accordingly, students are given a punched card at the beginning of the course which can be used as a template for coding alphabetic characters according to the Hollerith code. All of this answering is designed to be done in the comfort and leisure of the student's home rather than in the pressured atmosphere of classroom testing. If your students have free access to keypunches, they can readily punch their answer cards directly in the format for grading; they need only an ability to read the letters on the nearly standard typing keyboard of the keypunch to accomplish this. Of course, they must check to be sure the answers start and end in the correct card columns, a not impossible task when they are working at their own pace at home without classroom pressure.

## A. Types of Questions and Answers

### 1. General

The objective is to develop enough questions in the library to allow individualized sets of homework for each student, avoiding duplication of questions between students.

Any type of question which can be typed using normal keyboard symbols is amenable to application in this system. Answers can be in the same form, usually a word or a series of numbers of combinations of both, as will be shown later. Theoretically, the questions can be of any length, using a punch card for each line of print, and maintaining its identification number constant for as many cards as required. Answers may be of variable length (and therefore credit) in my latest program version, limited only to a maximum of nine characters per question card, or about 12% of the length of the question. If more length is required for either the answer or question, additional cards are used for whichever is desired, leaving the other portion blank.

With proper planning, questions and answers could be found for any subject matter, English literature, biology, economics, politics, history, as well as the physical sciences and mathematics.

In a sense, the answers become a form of one-dimensional crossword puzzle with all of the crossword's intellectual stimulation plus a self-checking feature, intriguing the student to find the word answer which fits the answer length asked for. One must be on the lookout for the possibility of more than one word of the same length having the same meaning. Often this is discovered by trial and error; since students are not bashful about informing the teacher of such a situation, upward adjustments of their credit will be required.

One possible method of writing questions to answer is to keypunch references to specific questions or problems in assigned texts on reserve at your local library. This has the advantage of exposing the students to additional literature on the subject, as well as offering them different approaches to the topic.

### 2. Examples of Specific Questions

Following are examples of questions from various disciplines which typify the scope of this system:

What name is given to the grammatical part of speech denoting action?
What year did Columbus discover the new world?
What great early philosopher wrote The Republic?
What year marks the Declaration of Independence?
Who first introduced mass production techniques into automotive manufacture?
What is the name of a plant structure containing nuclear and cytoplasmic material enclosed within a membrane?
What famous scientist won his second Nobel Prize for his efforts toward peace, a field unrelated to his first achievement?
How many cubic inches are there in a cubic foot?
Write the formula for hydrogen peroxide.
Write the coefficients for the chemical equation: $Fe_3O_4 + H_2 = Fe + H_2O$
Who wrote Pygmalion, also known as My Fair Lady?
What was the size of the Hiroshima and Nagasaki nuclear bombs (give number and unit abbreviation)?
Who made Stratford-on-Avon famous?
What is the answer to problem 153, page 8, Chapter One of Sackheim?
Calculate the number of inches in a meter, followed by the number of significant digits to the left of the decimal point.
Who is credited with revolutionizing the communication arts with the invention of the printing press?
What are the initials of the former President of Harvard whose Classics became known as the five-foot shelf?

## 3. Chemical Question Adaptations

Many chemical questions take the form of problems for which numerical solutions are required. These are the most readily adaptable forms to punched card operation, as only a single mark or punch need be made for any one card column answer. Some arbitrary decisions need to be considered in recording numerical answers on the cards. I shall cite those which I have adopted as examples, not necessarily indicating their use best for your own circumstances.

a. Short Integer Answers. Short integer answers are typified by a question such as: What is the number of protons in the nucleus of the sodium atom? One or two digit answers of this type are encoded as 11, just as might be anticipated.

b. Coefficients for Balancing Equations. Coefficients asked for balancing equations are to be encoded consecutively in the order in which the chemical components appear in the equation in question. If the student is expected to complete the equation before balancing, the use of a + sign and a blank space before the period is recommended. All coefficients, including ones (1) should be encoded, not assumed. This eliminates the possibility of shifting from the position locations of the correct answer.

For example, the coefficient answers for the following equation to be completed and balanced would be encoded as 1434:

$$Fe_3O_4 + H_2 = Fe + \quad .$$

The student would be expected to determine that $H_2O$ is the appropriate additional product expected and that its coefficient would be 4, the last digit of the answer indicated above.

c. Chemical Formulas. (1) Inorganic. The formula of the iron ore, magnetite, would be encoded as $Fe_3O_4$ and would print out on the grade sheet as all capital letters without subscripting, as FE3O4. Although parentheses are print characters available on data processing printers, their encoding involves a triple punch per column on the card and I prefer to have my students use the dash--or 11 punch instead, as it involves only a single punch. Thus, the formula for $Al_2(SO_4)_3$ would appear as AL2—SO4—3 from the printer.

(2). Organic. Organic formulas can be encoded in various forms, and some of the simpler examples will be discussed. The formula of diethyl ether might be expressed as the eight-character form $(C_2H_5)_2O$, which with dash encoding would print as —C2H5—2O. It could also be encoded as the nine-character form $C_2H_5OC_2H_5$ which would print as C2H5OC2H5. The letter, O, is encoded as the double punches, 11 and 6, which are readily distinguished from the single zero punch, although some printers print similarly appearing characters for both. Other printers print a squarish letter in contrast to the oval zero.

The numbering of the substituents of isooctane, used as an antiknock standard for gasoline, would be encoded as the three digits 224. Substituents should be numbered in order of increasing complexity, i.e., atomic or molecular weight of the substituent. Thus, methyl would precede ethyl, which would precede propyl, etc. The following compound would be encoded by substituent locations as the three digits 243:

$$CH_3CH(CH_3)CH(C_3H_7)CH(C_2H_5)CH_2CH_2CH_3$$

which represents the nomenclature of 2-methyl-4-ethyl-3-propyl heptane.

The formula for stearic acid could be encoded as the eight characters $C_{18}H_{36}O_2$, or the ten characters $C_{17}H_{35}COOH$, to demonstrate the functional character of the acid group. Since the latter exceeds the size of the answer field on the question card, it could be divided onto two successive cards into a six-character and a four-character portion answer, $C_{17}H_{35}$ and COOH, respectively, separating the aliphatic chain of the compound from its functional group.

(3) Formulas from Composition. Formulas from composition (percentage analysis or otherwise) should be encoded using the symbols for the elements in the same order in which they appear in the question, interspersed with numbers for subscripts greater than unity, as in regular formula fashion, $H_2O$ rather than $H_2O_1$.

d. Decimal Numbers. Decimals may be encoded using a triple punch in the same column. I prefer to avoid having the student encode the decimal point in an answer both because of the error possibilities involved in triple punching a card column, and also to avoid the loss of credit for both positions, which would ensue by virtue of a missplaced decimal, namely, the decimal and the digit it displaced. Therefore, to limit the credit involving placement of the decimal to one position (in most cases), I have the student encode the answer without decimal, starting with the first significant, nonzero, digit and proceeding normally, but reserving the last position of the answer (right-hand end) for a magnitude or integer count digit.

(1) Numbers Greater Than One. For numbers greater than one, the magnitude or integer count is obtained by counting all the integers to the left of the decimal point when the number is expressed in normal or nonexponential form. Numbers containing exponential notation should be converted to normal form, either mentally or actually. Conversion to exponential form with placement of the decimal point preceding all significant digits of the number followed by a base ten exponent makes the exponent equal to the integer count. Exponents, such as in Avogadro's number, require using the last two positions for the integer count. Table 1 gives a number of examples of numerical answers and their encoding, including normal rounding procedures.

TABLE 1

Encoding of Numbers Greater Than 1

| Number | Positions | Encoded | Printout |
|---|---|---|---|
| 1.23456 | 6 | 123461 | 123461 |
| 1.23456 | 4 | 1231 | 1231 |
| 123.456 | 5 | 12353 | 12353 |
| $6.0225 \times 10^{23}$ | 7 | 6022524 | 6022524 |
| $.60225 \times 10^{24}$ | 7 | 6022524 | 6022524 |
| .60225E + 24 | 6 | 602324 | 602324 |
| 10,000. | 4 | 1005 | 1005 |

(2) Numbers Less Than One. For numbers less than 1, the magnitude is determined by counting the number of leading zeroes between the decimal point and the first significant digit when the number is expressed in normal or nonexponential form. Leading zeroes are only those which determine the decimal placement or magnitude of the number and are not embedded within the significant digit's portion of the number. Conversion of the number to the normalized exponential form in which the decimal point is placed immediately in front of the first significant digit and ending in base ten exponential form makes the exponent equal to the number of leading zeroes found in the nonexponential form.

The encoding of the leading zero count is differentiated from the integer count by punching the zero in the same column with the count, thus making a double punch resulting in the printing of a character from the last third of the alphabet on the computer graded printout.

Usually only the last answer position is required for the leading zero count, but a number like the charge in coulombs on a proton, or its mass in grams, would require two positions.

Table 2 gives several examples, using a superscript and subscript to denote multiple punch encoding per position or card column. (See Fig. 9.)

e. Negative and Positively Signed Numbers. Although unsigned numbers are conventionally assumed to be positive, students are to prone to make errors in signs. E.g., for exothermic versus endothermic heats of reaction,

TABLE 2

Encoding of Numbers Less Than 1

| Number | Positions | Encoded | Printout |
|---|---|---|---|
| .123456 | 6 | 123460 | 123460 |
| .123456 | 5 | 12350 | 12350 |
| .123456 | 4 | 1230 | 1230 |
| .012345 | 4 | $123^{0}_{1}$ | 123/ |
| $.12345 \times 10^{-1}$ | 6 | $12346^{0}_{1}$ | 12346/ |
| .00123456 | 6 | $12346^{0}_{2}$ | 12346S |
| $.12345 \times 10^{-2}$ | 4 | $123^{0}_{2}$ | 123S |
| $1.60 \times 10^{-19}$ | 4 | $16^{00}_{18}$ | 16/Y |
| $.160 \times 10^{-18}$ | 5 | $160^{00}_{18}$ | 160/Y |
| $1.67 \times 10^{-24}$ | 5 | $167^{00}_{23}$ | 167ST |
| .167E-23 | 4 | $17^{00}_{23}$ | 17ST |

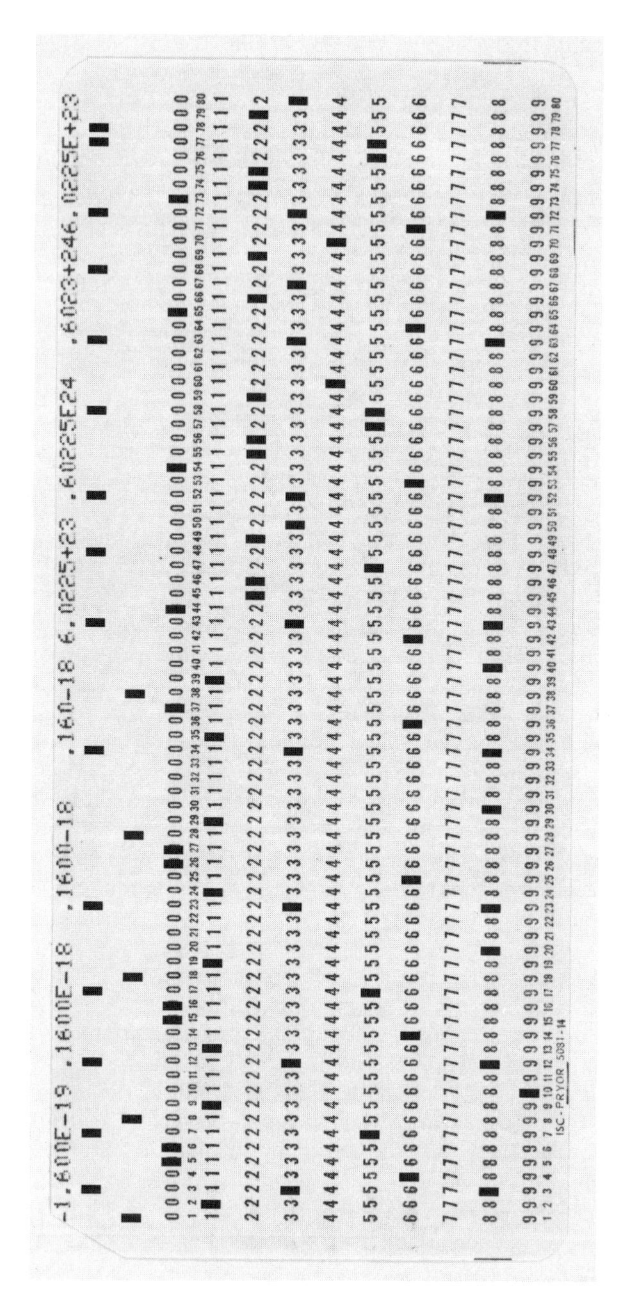


Fig. 9. General purpose card with data in exponential format.

the omission of a sign, falsely assumed to be positive when in actuality it is negative, results in displacement of the answer by one position throughout its entire length, and could lead to no credit for the answer when compared column by column with the key by the computer or by the manual sight method using an overlay. Since I like to give partial credit based on the number of digit or character positions matching the key, I have adopted the convention of always using the sign, + (a 12 punch) or - (an 11 punch) for any quantities which require it, such as voltages, heats of reactions, valences, work, entropy, temperatures other than Kelvin, and other thermodynamic quantities. This means that valences would commonly be two-position (two points credit) answers unless you preferred to dispense with signs completely and make valences single-position answers.

f. True, False, and Multiple-Choice. I commonly use the + (a 12 punch) for true and the - (an 11 punch) for false answers. Plusses and minuses are also useful for representing increase or decrease choices, along with the use of the zero for negligible change. An application of this latter situation is with Le Chatelier principle questions involving temperature, pressure, and other effects upon chemical equilibria.

Your present multiple-choice questions may be used in tests in the usual fashion, but the answers are recorded on punched cards for automated grading and record keeping of grades. This is a quick way of utilizing part of the versatility of this system while preparing your present questions for keypunching and your own "Custom Homework Library" use.

The zero through nine punches allow an expansion to a maximum of ten multiple choices, in contrast to the conventional five choices. These digit punches may be used for cross-matching tables, and enlarged to include the alphabetic characters for tables containing more than ten items to be matched.

As you see, multiple-choice questions are not precluded in this system. It was the inequity of lack of variable credit for problems more difficult and/or complicated problems than many nonmathematical multiple-choice questions that led to the development of this system, initially for testing, and subsequently for individualized "Custom Homework." Additional benefits were the elimination of the guessing feature of multiple-choice questions, and, especially the elimination of similar or common multiple answers.

## B. Format of the Question Cards

### 1. Answers and Length

The first nine card columns (cc 1 to 9) are arbitrarily reserved for the answer. An answer is required in the first card and may overflow into the same field (cc 1 to 9) of succeeding cards. The answer should be left justified, i.e., it should begin in cc 1 and proceed as you would normally write from left to right, until it is completed within the first nine card columns of each card. Any number, letter, or special character found on the keyboard may normally be used as a part of the answer, provided each student has a copy of the required coding shown in Fig. 1. I usually hand out one of these cards to each student at the beginning of the semester.

Answers exceeding nine characters in length may overflow onto succeeding answer fields, split in any desired fashion. For example, an answer like "Shakespeare" might be split into its syllables, with five positions on the first and six on the second card. Such a system aids the student also, because it confirms the length of the correct answer. It is possible to tell the student within the text of the question that the answer is split into two syllables, or into parts with two syllables in the first part, and one syllable in the second part, etc. Again, the computer numbers the questions to reflect the length of answer or partial answer entirely at the question designer's discretion. For example, if you wish to value a question at only two points, you could ask only for Shakespeare's initials, "WS." In other words, you are limited only by your ingenuity in phrasing your question to elicit the number of answer positions you desire. If you so desired, you could ask the student for the first, ninth, and seventh letters of the bard's last name in that order, for further variation in the answer length and, therefore, credit value. The examples given are intended to give some idea of the realm of possibilities with this system, rather than indicating limitations of the system.

The number of answer positions actually utilized in each card is punched as a single digit into card column 13 of the same card in which the answer appears. This controls the incrementing of the question numbering on the "Custom Homework" printout as well as positioning the answers contiguously on the keycard punched by the computer. The question numbering on the "Custom Homework" printout tells the student how many positions are allotted for each answer on the student answer card.

If more cards are required to express the body of the question than are required for the answer, the answer fields (cc 1 to 9) are left blank, and card column 13 may be left blank. It may contain a zero, - or + punch, but may not contain any digits 1 through 9 unless there is an answer in that card related to such length. The blank, zero, - or + punch could be utilized for machine sorting in case the cards should get out of sequence.

## 2. Question Classification and Identification

Classification of the question according to topical material is accomplished in cc 10. This normally allows ten decimal subdivisions of the material, but this may be expanded to 36 subdivisions by overpunching with a + or - punch. I retain ten basic subdivisions, but expand the number of questions by using A, J, and / which result from overpunching a +, -, or a zero in the same card column as the digit 1 is punched. In numerical mode, these letters all sort mechanically into the same pocket (1) of the sorter. Similarly, 2, B, K, and S all sort into the 2 pocket.

The question number identification is continued into cc 11 and 12. Thus, all cards for the same question should have the identical characters and numbers punched into cc 10 through 12, so that the computer or 407 machine can recognize and number each question in the printout, double-space for new questions, etc.

Card column 14 is reserved for the ease decile based on the proportion of students answering the same question correctly on exams. This would allow you to pick the range of ease in compiling future exams or homework sets. Alternatively, it would give you a measure of the ease of each set or exam for comparison with different sections in the same or previous years.

## 3. The Text of the Question

The text of the question may be expressed using any of the standard numbers, letters, or special characters available on the keyboard of a keypunch machine compatible with your computer. Some computers accept more (and sometimes different) multiple punch combinations in the same card column than others, or they may have different character-set print symbols for these special punch combinations. Some printers, for example, may print a pound or number sign for one parenthesis, and a lozenge for the other. If you plan to use special characters, it is best to punch trial cards and print them on your particular equipment and view the results.

It is important to bear in mind that most computers do not handle lowercase (noncapital) letters, so that chemical symbols in formulas always come out printed entirely in capital letters. Additionally, subscripts and superscripts are not differentiated from ordinary digits, because all are printed directly in line horizontally without being raised or lowered. Underlining is not usually available, so it is difficult to construct the usual "over and under" numerator and denominator fractions, either in chemical notation, as in equilibrium constants, or conventional mathematical notation. Instead, the slash / usually denotes division, assisted by liberally used parentheses to avoid error or misplacement of denominators, etc.

Structural organic formulas are best written in linear form as:

CH3-CH(CH3)-CH(CH2CH3)-CH(CH(CH3)2)-CH2-CH2-CH3

or:

CH3CH(CH3)CH(CH2CH3)CH(CH(CH3)2)CH2CH2CH3

rather than:

```
                H
                1
              H-C-H
                1
      H     H H-C-H H     H     H     H
      1     1   1   1     1     1     1
    H-C  -  C - C - C  -  C  -  C  -  C-H
      1     1   1   1  H  H     1     1
      H   H-C-H H H-C-C-H       H     H
            1       1  H
            H     H-C-H
                    H
```

or other combinations which are time-consuming to construct on punched cards. Two-dimensional constrictions are found to be even more severe on paper than steric hindrances in three-dimensional space. Multiple bonds may be depicted using an equal sign for double CH2=CH2, and both a dash and an equal sign for triple CH=-CH.

Card columns 15 through 80 are utilized for the text portion of the questions, with as many cards as needed--theoretically there is no limit. Each successive card on the same question should have the same three-character identification in cc 10-12, as indicated in Section IV. B. 2.

## V. "CUSTOM HOMEWORK" ASSEMBLY

The "Custom Homework" section of this system is integrally related to the library of questions and answers. As the term "custom" implies, each student will receive a unique set of questions maximizing individual effort (and reward) in a manner heretofore impractical from a grading time standpoint.

Although the question cards (with answers) are processed in this system in whatever order they are fed to the computer, no sequencing results; in fact, mine are always well-mixed from insertions, deletions, and mixing during the course of selections for exams, and some deliberate shuffling. Further randomization may be accomplished by sorting on one of the I. D. number card columns or by storage on a disk memory and calling questions by means of a random number generator program for your computer.

All three methods of "Custom Homework" assembly described here produce homework sets similar in principal to that shown in Table 5 (see p. 231) with minor variations in format. Each page of the printout is numbered sequentially, and each question is numbered sequentially with the interval between numbers equal to the number of characters required for the answer. The student therefore knows the length of the answer much as he would by counting the spaces in a crossword puzzle. The page numbering is printed in two locations, once on the question portion and again on the answer portion. The two portions are aligned vertically on the page for ease in comparing and checking answers visually or for manual grading.

Which method you choose to prepare "Custom Homework" depends on the availability of your equipment and the degree of integration of the various sections you wish to use. Some of the distinctions between the methods follow.

Use of the 407 accounting machine method can be a good way to learn the operation and allow you to practice much less expensively than immediately trying to adapt the system to your computer. As programmed here, the 407 only prints the questions and answers, without producing any punched keycards. This means the grading must be done manually by visual comparison of the student's written answer with the printed key answer. This is an advantage for both the students and the instructor during the initial learning and debugging phase of the system. Even without any special board wiring (which can be self-taught), you would merely use a standard 80-80 listing board and print the questions and answers on continuous perforated sheets, label them with student names or initials and easily grade them visually.

The RPG program described in this section is directly suitable only for IBM 360 computers with RPG compilers. Other computer manufacturers have similar languages. For these languages, the example given here is only indicative of the magnitude and style of programming. On the IBM 360 Model 20, this RPG program is limited to fixed-length answers because each answer is punched into a predetermined location on the keycard. The program can be altered to provide for a different fixed-length answer more convenient to the answer card.

The FORTRAN program described in this Section is the most universally applicable--all but the smallest computers usually have FORTRAN compilers. This is the most versatile of the three methods of "Custom Homework" assembly in that variable-length answers are feasible, punched in contiguous fashion on the keycard simultaneously with each page number of homework as a cross reference. The keycard along with the "Custom Homework" printout then becomes the machine-readable link between the homework answers and the computerized grading system described in Section VII.

Each student receives an answer card (if machine grading is desired) and the corresponding page of homework. The answer card contains the student name, number, and section and the page number of the homework. The answer cards may be machine-alphabetized for quick distribution before or after class. The answer cards are easily machine-produced from the keycards after machine punching the student identification into the keycards at random. Upon return of the student answer cards, they can be merged by machine with the keycards for machine grading.

## A. Printed by the 407 Machine

### 1. Paper Control and Card Feeding

The continuous paper (forms) commonly utilized by the 407 machine and computer printers is commonly perforated with every 8.5 or 11 inches and may be purchased in varying widths to meet the print-width capacity of the printer. Most printers range from 120 to 144 characters in width at ten characters per inch. For the application indicated here an 11 inch form would suffice, exclusive of the two half-inch margins which are punched for automatic advancement by the tractor wheel feed system.

Most installations should have prepared carriage control tapes to match the paper (form) sizes commonly used. Such tapes assure that printing of the heading, once adjusted, shall subsequently appear at the top of each successive form, followed by the questions. The location of this first print or heading line of each page is determined by punching a hole on the first line (channel 1) of the carriage control tape at the desired point. Skipping from the bottom of each page (allowing for a margin) to the next page's first desired print line is determined by punching the tape on the last line (channel 12) at the desired point of the last printline (See IBM Manual A24-1011-2, pp, 66-68) [6].

Each printline of a question is single-spaced with triple spacing between questions. An 8.5-inch form has about 45 usable printlines. An 11-inch form can be used for lengthier sets of homework. My usual assignment is one page per student.

The punched card deck of questions first has a header card with the day punched in cc 17 and 18, the month in cc 19, and the year of the decade in cc 20 and a dash - in cc 14 which signals the 407 that this is the header date card. The questions may be in any random order, and will be printed in the order they pass through the machine.

## B. Computer-Printed with Keycard Punching

In contrast to the wiring program used in the 407 machine, the computer is controlled by written programs (called software) giving it much more versatility. These programs automatically produce keycards for each page of homework in addition to printing the questions and the answers.

### 1. Fixed-Length Answers (RPG)

This initial program was written in the REPORT PROGRAM GENERATOR (RPG) language, which is oriented to IBM computers. The answer lengths were arbitrarily fixed at four positions each to readily handle usual slide rule accuracy of three significant digits plus an integer count in the last position.

a. Programming. The program is shown in Table 3 and is discussed in the following paragraphs referenced to the card numbers punched into the program cards in cc 1-4.

(1) Cards 0102-4 describe the files and devices which process them. The questions are placed in the hopper of the READer. The OUTPUT is obtained from the PRINTER, and the keycards from the PUNCH.

(2) Cards 0301-8 describe the INPUT header card containing the DAy, Month, and YR., in cc 61-64.

(3) Cards 0309-12 describe the question cards with ANSWER in cc 1-4, EASE decile in cc 14, and QUESTion in cc 15-80.

(4) Cards 0401-04095 describe the printed OUTPUT heading: "DUE DA M YR PAGE"

(5) Cards 0410-04185 describe the end positions for printing the values under the heading above.

(6) Cards 0419-04215 indicate the end positions for printing the QUESTNo., the QUESTion, and the ANSWER.

(7) Cards 0430-44 describe the KEYCARD format punched with PAGe number, and the seven four-position answers ending in cc 58. The EASEAVerage is punched in cc 30.

(8) Cards 05012-0524 contain the various calculation instructions and operations for setting up the ANSWER and QUESTNumber locations in memory.

Further programming and processing details may be obtained from IBM File No. S360 (Mod 20) - 28, Form C26-3600-5 [7] or related publications.

b. Forms Control. Positioning and advancing of the paper (forms is controlled by a carriage control tape as described previously (see Section V.A.1.). Shorter (in depth) forms are required since each of the seven printlines is considered a new question number incremented by four, and four positions are placed on the keycard.

c. Keycards. The keycards obtained from the punch unit have the first nineteen card columns reserved for subsequent student number and name, and cc 24-29 for course identification. The page number of "Custom Homework" used is punched into the keycard in cc 20 to 23 so the student can select the correct homework page matching his answer card. The seven four-position answers are punched into the 28 card columns from cc 31 through 58. The student number, name, and course identification is reproduced into the keycards from cc 1-19 and 24-29 of the master deck (see Fig. 2) using a single-card reproduce program of the computer, or the IBM 514 Reproducer described (in Section III. E), or, as a last resort, partial reproduction in a keypunch. Alternatively, a duplicate deck of the master cards may be used as input for the computer punch unit, although punch verification cannot be made in this instance.

I find it convenient to duplicate I. D. information on the keycards afterward with advancement of the page number of the homework received by each student by switching the last student master card to the front of the master deck each time I process them. This also rotates the selection to even out any bias toward easier questions at the beginning of each set. It also time-sequences the key and answer cards in the absence of specific date punches therein. A + punch is added to cc 30 of each keycard using a single card gangpunch program for the computer, or the emitter unit of the IBM 514 Reproducer, or it can be done on a keypunch. This + (or 12) punch serves to differentiate the keycards from the student answer cards during grading.

d. Student Answer Cards. (1) Mark Sense Cards. Mark sense student answer cards are prepared by partial duplication of the keycards using only cc 1-29 containing the identifications but not the answers. These can be duplicated readily by computer, reproducer, or even by keypunch, into the mark sense cards shown in Figs. 3 and 4.

(2) Port-a-punch Cards. Partial duplication of cc 1-29 of the keycards into the Port-a-punch cards is somewhat different than with the mark sense cards because the information must be transferred to odd-numbered card columns. This is accomplished in the IBM 514 Reproducer, using a wiring board. Keycard cc 1 through 29 are read and punched into the Port-a-punch answer cc 1, 3, 5, 7, 9, through cc 57, leaving the even-numbered cc 2 through 80 for student answers. Port-a-punch cards are shown in Figs. 6 and 7. The even-numbered card columns ONLY of Port-a-punch cards are specially pre-perforated for easy removal with a pen or pencil point by the student anywhere. The odd-numbered card columns can only be punched with typical keypunch equipment.

Again this conversion can be accomplished by programming your computer. If both machines are unavailable, it would be wise to prepare your master list using only the odd-numbered card columns, which could then be quickly reproduced into the Port-a-punch answer cards in a few seconds. Occasionally I have found it handy to have master cards in this latter format readily available for quick 80-80 card column reproduction into Port-a-punch answer cards for tests or quizzes not involving homework page numbers, using any equipment available at the moment.

e. Alternate Version in RPG. An alternate version prints six homework problems per page, punching the keycard answers into cc 32-55. Answers for problems numbered 2 through 6 appear in the same card columns as required for the program on percentage grading of numerical results described later (in Section VI. A. 1. a.).

2. Variable-Length Answers (FORTRAN)

The major advantages of variable-length answers lies in permitting credit proportional to answer length, and at the same time providing the student with the self-checking feature of a one-dimensional crossword puzzle.

FORTRAN was chosen as the programming language because it is so widely used. Version II was used as it is almost universally compatible with computers using later versions. Later versions, such as FORTRAN IV, are more flexible, but lack compatibility with older computers.

a. Programming. The program is shown in Table 4 and will be discussed in sequence using the left margin statement numbers as references where given, and capitalizing here, program words used.

TABLE 3

Computer Program in RP6 for "Custom Homework"

```
0000  *  -YANEY   07FEB70 VERSION HOMEWORK                          0720
0102 FQUESTIONIPE                           READ01
0103 FOUTPUT  OS                            PRINTER
0104 FKEYCARD OS                            PUNCH42
0301 IQUESTIONAA  99   10 Z-
0306 I                                            61   620DA
0307 I                                            63   63 M
0308 I                                            64   640YR
0309 IQUESTIONAB  01   10NZ-
0310 I                                             1    4 ANSWER
0315 I                                            15   80 QUEST
0312 I                                            14   140EASE
0401 OOUTPUT   H  1 1    1P
0402 O         OR        L1
0406 O                                        35 'DUE'
0407 O                                        40 'DA'
0408 O                                        45 'M'
0409 O                                        50 'YR'
040950                                        60 'PAGE'
0410 OOUTPUT   D  1      99
041050         OR        L1
0411 O                           PAGE  Z      60
041150                           PG    Z       4
0415 O                           DA           40
0417 O                           M            45
0418 O                           YR           50
```

TABLE 3 - Continued

```
041850                         ANSWER     4
0419 0        D 11    02NL1
0420 0                         QUESTN    12
0421 0                         QUEST     80
042150                         ANSWER     4
0430 OKEYCARD D        L1
0433 0                         PG    Z   23
0438 0                         A1        34
0439 0                         A5        38
0440 0                         A9        42
0441 0                         A13       46
0442 0                         A17       50
0443 0                         A21       54
044350                         A25       58
0444 0                         EASEAV    30
05012C                     MOVE PAGE      PG      30
05014C                     MOVE 01        EASEA   20
05016C   01      COUNT     ADD  1         COUNT   20
05017C   01      COUNT     COMP 1                        89
05018C   01      EASE      ADD  EASEA     EASEA   20
0502 C   89                MOVELANSWER    A1      4
0503 C   89                MOVE '01'      QUESTN  4
05035C   89                SETON                      02
0504 C   01      COUNT     COMP 2                        88
```

TABLE 3 - Continued

```
0505 C    88                  MOVELANSWER   A5       4
05055C    88                  MOVE '05'     QUESTN   4
05056C    88                  SETON                     02
0506 C    01       COUNT      COMP 3                          87
0507 C    87                  MOVELANSWER   A9       4
0508 C    87                  MOVE '09'     QUESTN
05085C    87                  SETON                     02
0509 C    01       COUNT      COMP 4                          86
0510 C    86                  MOVELANSWER   A13      4
0511 C    86                  MOVE '13'     QUESTN   4
05115C    86                  SETON                     02
0512 C    01       COUNT      COMP 5                          85
0513 C    85                  MOVELANSWER   A17      4
0514 C    85                  MOVE '17'     QUESTN   4
05144C    85                  SETON                     02
0515 C    01       COUNT      COMP 6                          84
0516 C    84                  MOVELANSWER   A21      4
0517 C    84                  MOVE '21'     QUESTN   4
0518 C    84                  SETON                     02
05182C    01       COUNT      COMP 7                          83
05184C    83                  MOVELANSWER   A25      4
05186C    83                  MOVE '25'     QUESTN   4
05188C    83                  SETON                     02
0520 C    83       EASEA      DIV  COUNT    EASEAV  104
05202C    02       COUNT      COMP 8                          L1
0523 C    L1                  MOVE 0000     COUNT
0524 C    L1                  SETON                     89
```

TABLE 4

Computer Program in FORTRAN for "Custom Homework"

```
*LDISKHOMWRK
      DIMENSION A(9)
      DIMENSION B(50)
      J = 1
      I = 1
   90 FORMAT(19X,I4,7X,25A1,25A1)
    1 FORMAT(1H1,31X,3HDUE,3X,2HDA,4X,1HM,3X,2HYR,6X,4HPAGE)
    2 FORMAT(16X,I2,A1,I1)
    3 FORMAT(9A1,A3,A1,A1,16A4,A2)
    4 FORMAT(/1X,9A1,    A3,4X,I2,2X,16A4,A2)
    5 FORMAT(1X,I2,1X,A1,1X,I1,1X,I4,26X,I2,4X,A1,4X,I1,6X,I4,/)
    6 FORMAT(21X,16A4,A2)
      ICNT = 0
      IPAGE = 0
      IPOSTN = 1
   22 READ 2,IDAY,MONTH,IYEAR
      IPAGE = IPAGE + 1
      PRINT 1
      PRINT 5,IDAY,MONTH,IYEAR,IPAGE,IDAY,MONTH,IYEAR,IPAGE
   14 READ 3,A,AD,ILENTH,ICON,QONE,QTWO,QTHR,QFR,QFV,QSX,QSV,QEG,QNN,QTN
     1,QELV,QTWLVE,QTHRTN,QFRTN,QFIFTN,QSXTN,QSVTN
      IF(7100 - ILENTH)10,10,11
   10 ILENTH = (ILENTH-7000)/100
      IF((ICNT+ILENTH)-50)12,12,13
   12 ICNT = ICNT + ILENTH
      PRINT 4,A,AD,IPOSTN,QONE,QTWO,QTHR,QFR,QFV,QSX,QSV,QEG,QNN,QTN,QEL
     2V,QTWLVE,QTHRTN,QFRTN,QFIFTN,QSXTN,QSVTN
```

TABLE 4 - Continued

```
      IPOSTN = IPOSTN + ILENTH
  100 B(I) = A(J)
      IF(J - ILENTH)98,99,99
   98 J = J+1
      I = I+1
      GO TO 100
   99 I = I + 1
      J = 1
      GO TO 14
   13 PUNCH 90,IPAGE,(B(N),N=1,ICNT)
      J = 1
      I = 1
      IPOSTN = 1
      ICNT = ILENTH
      IPAGE = IPAGE + 1
      PRINT 1
      PRINT 5,IDAY,MONTH,IYEAR,IPAGE,IDAY,MONTH,IYEAR,IPAGE
      PRINT 4,A,AD,IPOSTN,QONE,QTWO,QTHR,QFR,QFV,QSX,QSV,QEG,QNN,QTN,QEL
     2V,QTWLVE,QTHRTN,QFRTN,QFIFTN,QSXTN,QSVTN
      IPOSTN = IPOSTN + ILENTH
      GO TO 100
   11 PRINT 6,QONE,QTWO,QTHR,QFR,QFV,QSX,QSV,QEG,QNN,QTN,QELV,QTWLVE,QTH
     1RTN,QFRTN,QFIFTN,QSXTN,QSVTN
      GO TO 14
    9 CALL EXIT
      END
```

(1) The initial card listed is one of the control cards for the particular computer used.

(2) The DIMENSION statements set up the core memory for each nine-position answer A, and the 50 answer positions B for the keycard. The next statements initialize program steps.

(3) Statements 90 and 1 through 6, give the sequential FORMATs to be READ, PRINTed or PUNCHed by later commands. Number 90 is for the keycard. Number 1 is the heading for each page, and number 5 is for the values to be printed under each heading. Number 2 is for reading the date card, and number 3, the question cards. Number 4 and 6 represent the first and succeeding printlines of any question, and the answer and other identification appearing only with the first line. Three more initializer statements follow.

(4) Number 22 READS the date card, the PAGE is incremented, the header is PRINTed followed by its appropriate values.

(5) Number 14 READs the question card while various checks are made to see if enough answers will be accumulated to exceed the 50 positions available on the keycard [to just beyond statement (10)]. If not exceeded, the question is printed.

(6) Statement 100 and those following move the appropriate answer positions into the 50 positions of core storage until filled, after which the key card in PUNCHed according to statement 13.

(7) The remaining statements reset conditions for PRINTing the new page headings and then PRINTs the question line which had been held during the PUNCHing operation, after which control is looped back to statement 14 for READing the next question.

b. Homework Printout, Key, and Answer Cards. A sample "Custom Homework" printout is shown in Table 5, using a control tape for a short form so that part of page 2 of the homework can be seen following that of page 1. Normally the control tape would be punched to advance page 2 to the next sheet as previously described (see Section V.A.1.).

The keycard produced for page 1 of the Homework printout is shown in Fig. 10. Student identification is reproduced into the vacant columns at the beginning of the card in the same manner as described previously (see Section V.B.1.c). The student answer cards are then prepared by partial reproduction (see V.B.1.d).

## VI. GRADING BY MACHINE

The underlying principle involved in the main program of this section and its modifications is a comparison of the student answer cards with the keycard(s) and counting the number of matching character punches. Each matching character counts one point (or a percentage of the total points).

TABLE 5

Custom Homework with Variable Length Answers (FORTRAN)

```
                                   DUE    DA     M    YR        PAGE
 3 J 1     1                               3     J     1          1

1492       002     1   WHAT YEAR DID COLUMBUS DISCOVER THE NEW WORLD

PLATO      003     5   WHAT GREAT EARLY PHILOSOPHER AUTHORED THE REPUBLIC

1776       004    10   WHAT YEAR MARKS THE DECLARATION OF INDEPENDENCE

PAULING    005    14   WHAT FAMOUS SCIENTIST WON HIS SECOND NOBEL PRIZE FOR PEACE

20KT       006    21   WHAT WAS THE SIZE OF THE HIROSHIMA AND NAGASAKI NUCLEAR BOMBS

SHAW       007    25   WHO AUTHORED PYGMALION-ALSO KNOWN AS MY FAIR LADY

3660       008    29   WHAT IS THE ANSWER TO PROBLEM 153-PAGE 8-CHAPTER I OF SACKHEIM

SHAKESPEA009      33   WHO MADE STRATFORD ON AVON FAMOUS

1728       000    42   HOW MANY CUBIC INCHES ARE IN A CUBIC FOOT

FORD       011    46   WHO FIRST INTRODUCED MASS PRODUCTION TECHNIQUES INTO MANUFACTURING
                                   DUE    DA     M    YR        PAGE
 3 J 1     2                               3     J     1          2

CELL       012     1   WHAT IS THE NAME GIVEN TO A PLANT OR ANIMAL UNIT OF STRUCTURE
                       WITH NUCLEAR AND CYTOPLASMIC MATERIAL ENCLOSED WITHIN A MEMBRANE

602323     013     5   WRITE THE FIRST FOUR DIGITS OF AVOGADROS NUMBER WITHOUT THE DECI-
                       MAL POINT FOLLOWED BY THE DIGITS REPRESENTING THE EXPONENT OR POW-
                       ER OF 10 IF THE DECIMAL WERE IMMEDIATELY AFTER THE FIRST SIGNIFI-
                       CANT DIGIT OF THE NUMBER
                       THIS 5TH LINE PER QUESTION IS THE MAXIMUM FOR THIS PROGRAM.

NACL       014    11   WRITE THE FORMULA FOR ORDINARY TABLE SALT

39372      015    15   CALCULATE THE NUMBER OF INCHES IN 1 METER FOLLOWED BY THE NUMBER
                       OF SIGNIFICANT DIGITS LEFT OF THE DECIMAL POINT

MENDELEEF016      20   WHAT FAMOUS RUSSIAN SCIENTIST IS CREDITED WITH DEVELOPMENT OF THE
                       PERIODIC TABLE OF CHEMICAL ELEMENTS ESSENTIALLY IN ITS PRESENT
                       FORM

CH3CH2CH28        29   WRITE THE FORMULA FOR STEARIC ACID
```

One of the programs is limited to mathematical problems with numerical answers. It determines the percentage error of each of several answers and averages and subtracts from 100 for the grade.

All student answer cards are processed to convert them to the format of the keycard(s). Mark sense cards are processed through the reproducer (IBM 514) to convert the special pencil marks to punches in the same card. Port-a-punch cards are read by the reproducer (or a computer) and new answer cards produced in the proper format. Partially prepunched student answer cards can be distributed into which the student can punch in the proper format his entire answer selections, thus eliminating conversion.

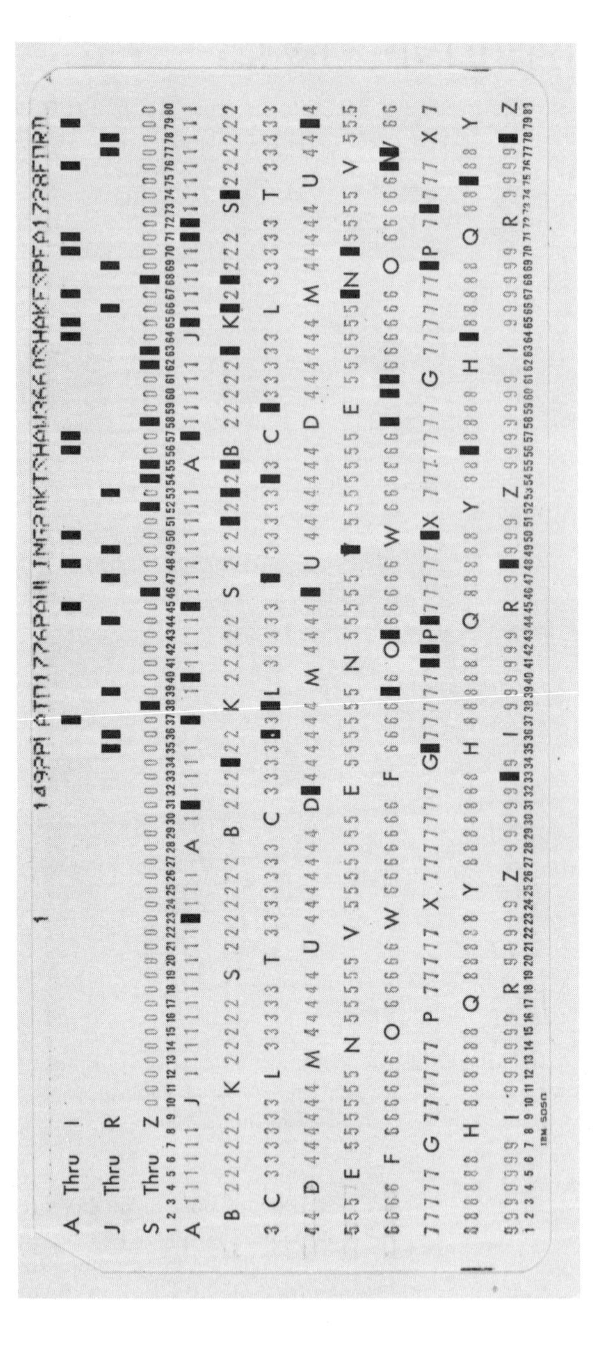


Fig. 10. Keycard from custom homework.

At this stage the answer cards may be visually compared with the keycard by placing the latter over each student answer card and marking in red all unpunched areas (errors) showing through (correct) punch holes in the keycard. These errors may then be counted for manual grading. Special optical scanners are available for reading mark sense cards directly by the computer.

For "Custom Homework" answer cards, merging of the keycards is necessary so that each keycard precedes its corresponding student answer card (Fig. 11). This can be accomplished with the collator or a sorter (some computers also have this capability in their hardware).

An alternative program will grade the conventional type quiz and test cards in which all students have common questions whose answers are all compared against a single common keycard.

Various auxiliary features are available following the complete printout of all keycard answers and student responses, with appropriate identification and student scores. These include a count of the number of students, total scores, average score, and mean and standard deviation. This information is also punched into blank cards for record-keeping and use in grade sub- and final totalling described in the next section. In addition, portions of the programs will give the grade frequency distribution by percentile, and analysis of each answer position for the number and/or percentage of correct responses.

Fig. 11. Collator, wiring board, merging answer cards with keycards.

Detailed descriptions of the procedures used are available from the author.

## A. Computer Grading

### 1. System 360 RPG Programs

a. Percentage Grading of Numerical Answers. This program gives proportional credit for numerical answers to homework problems printed according to the alternate version in RPG (see Section V. B.1.e). It can also be used for grading experimental laboratory results with single or replicate values.

This program was designed for use of the mark sense card shown in Fig. 3, but the data may be keypunched into cc 36-55 of the answer card instead.

Five answers containing four digits each are marked in Fields 1 and 2, 3 and 4, 5 and 6, 7 and 8, and 9 and 10, respectively, of the mark sense cards, or they may be punched in the contiguous card columns 36-55 indicated above.

The program compares each answer with the key value and computes the percent error to a maximum of 99%. The individual percentage errors are averaged and subtracted from 100 to give the % grade. The printout contains the usual identification, the answers and their individual percentage errors and the grade, followed by the total grade sum, count of students, and overall average grade. The printed information is also punched into blank summary cards for recordkeeping purposes.

b. Common Keycard Grading of Quizzes and Tests. This program is suitable for grading the conventional tests in which all students answer the same questions and are all compared against a single common key. However, it surpasses the conventional multiple-choice type tests, by offering both full alphabetic and all ten-digit numeric capability for answers, in conjunction with the usual five multiple-choice questions.

The header and keycard information is printed and each student answer card is compared with the keycard, and their answers are printed for checking. The number of correctly matching answers is printed on the same printline along with the student identification. The number of students, total grade, and average grade are printed at the end of the listing. All printed scores are also punched in summary cards for the record file.

c. "Custom" Keycard Grading. This program grades "Custom Homework" answer cards which have a separate and unique keycard for each student answer card. It is also applicable to grading of individualized laboratory unknowns, such as issued in qualitative or quantitative analysis from printouts of the keycards. The procedure is similar to that described above in that a header card is required in front of the deck to be graded, but in this case, the individual keycards must be merged with the student answer cards as described in Section VII. A. Each keycard is printed out immediately ahead of the student answer card with which it is compared. As with the previous program, the number of students, total grade, and average grade are printed at the end of the listing, and student score cards (Fig. 12) with the homework identification, etc., are punched for record-keeping.

### 2. FORTRAN Grading Programs

a. Common Keycard(s) for Quizzes and Tests. (1) With Question Analysis. This two-page program was written for a computer with moderate (20 K) core memory, in contrast to all those previously described which required a maximum of 8 K (many actually much less than 4 K) core memory. The keycard(s) answers are printed at the head of the report, followed by the student answers and scores (both pointwise and as a percentage), number of students, total of scores, total of squared scores, mean and standard deviation. The question analysis follows, giving the number of correct responses for each question.

Summary student score cards are punched for record-keeping purposes containing the contents of the first 72 card columns of their answer card, followed, in the last card, by the score in cc 75-77, and the percentage score in cc 78-80.

(2) With Grade Frequency Distribution and Question Analysis. This three-page program requires more core memory than the previous program, dependent to a considerable extent upon the number of students you program it to handle, in the DIMENSION statements (FORTRAN).

The first or header card contains the number of answer positions (max. 150) per test in cc 1-3, the number of answer cards (max. 3) per student in cc 4, the course number in cc 5-7, and codes in cc 8 and 9 for frequency grade distribution for each section and for the combined sections, respectively. A zero requests the frequency distribution, while a 1-punch suppresses. For example, there would be no need to print both if only a single section of students was being graded.

The sequenced keycards and sequenced answer cards for the students follow the header card in the deck with an end card containing 999999 in cc 1-6.

The keycard(s) print at the head of the report followed by the student answers and scores (in points and percentages), number of students, total of scores, total of squared scores, mean and standard deviations. The grade distribution(s) by frequency and percentile are printed next, if requested, followed by the question analysis, giving the number of correct responses for each question. Summary student score cards are punched for record-keeping purposes as in the preceding program.

b. "Custom" Keycard(s), with Grade Frequency Distribution. This two-page program was processed by a moderate size (20 K) core memory computer. It gives the number of scores and the average, the total of scores, total of squared scores, mean and standard deviation, as well as the grade distribution by section with the frequency of each score and its percentile rating. No question analysis was programmed, as it would be meaningless due to the diversity of the questions found in "Custom Homework" or laboratory unknowns issued.

A maximum of three pairs of key and answer cards per student may be merged or collated and graded within a deck and summed to give a single score per student. Normally, homework assignments or lab unknowns would be on single cards. Another use is multicard take-home exams. The program will, of course, grade single pairs of keycards and answer cards.

The first or header card contains the number of answers to be graded on the first keycard in cc 1 and 2, the number of answers on the second keycard in cc 3 and 4, and the number of answers on the third keycard in cc 5 and 6. Each student answer card should be preceded by the right keycard, although answer cards may be missing. As in the previous program, summary student score cards are punched during printing of the report.

## VII. GRADE SUB- AND FINAL TOTALING

### A. Preliminary Processing

The summary student score cards (Fig. 12) produced for record-keeping during the computer grading and/or those punched manually need to be merged or collated into a single deck in which all score cards for each student are grouped together for processing. The students need not be arranged alphabetically, although this is usually desirable.

### B. Manual Grade Summation

Using the merged decks (as done in Section VII.A), a complete listing is readily made by passing them through the 407 machine using a standard 80-80 listing board supplied, or passing them through any computer preceded

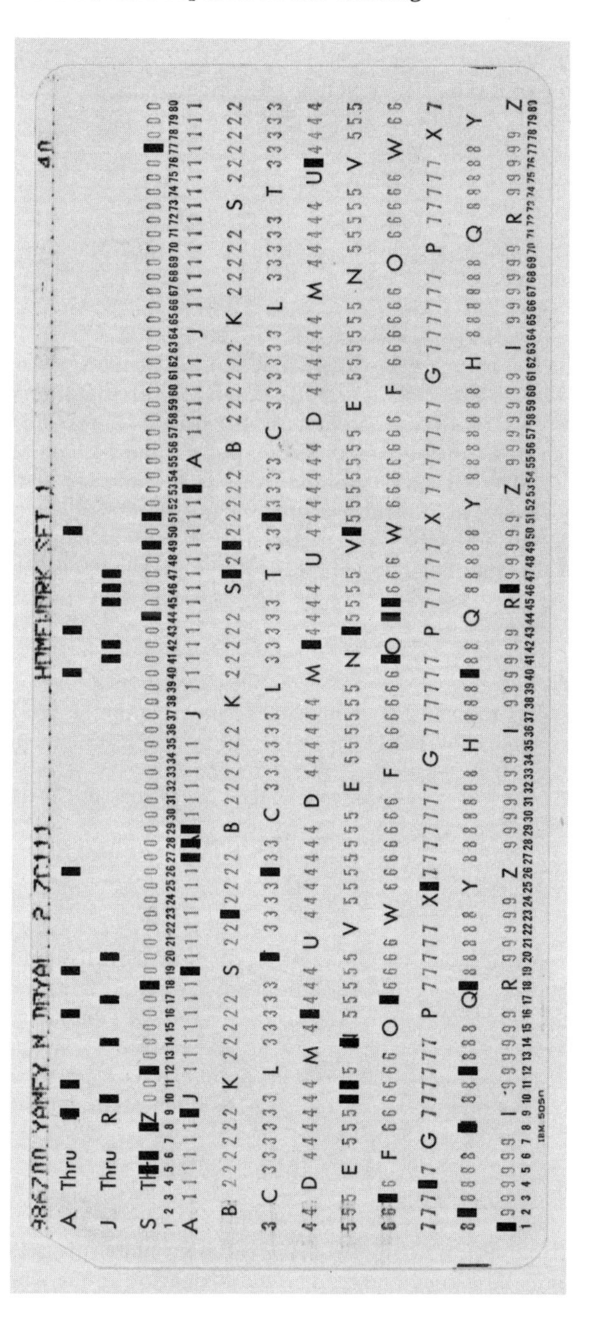

Fig. 12. Student score card.

by a single (usually) card 80-80 read and list program. This lists all information on the cards including the student scores which can be readily added manually or with a calculator. The same can be done from the cards themselves, although neither of these take full advantage of the investment you have made in this system.

## C. Grade Summation by IBM 407 Machine

### 1. Single Score per Card

Wiring a board to read, print, and add the score from each card and give a total of scores for each student is relatively straightforward. The reader is referred to IBM Reference Manual (A24-1011-2) for the description of wiring diagrams for detail printing on pp. 13 and 14, for MINor PROGRAM START & COUNTER ENTRY on pp. 15 and 16, and for addition on pp. 117 to 120. A change in student number (cc 1-6) is considered a minor change and is wired to the comparing units on first and second reading, and causes printing of the summary score total for the previous student.

### 2. Multiple Scores per Card

Wiring a board to crossfoot 11 scores from the same card, printing and giving the total for each student card is more complicated than for a single score per card. The reader is referred to pp. 150 through 153 of the IBM manual.

## D. Grade Summation by Computer

### 1. System 360 RPG Program

This program sums single scores per card for all the cards of a student, printing the entire contents of each card as read, giving a count of the cards per student, the total points and the average. A summary student score card containing student number, name and course identification, the number of scores, the average, and the total points to date is punched out for each student. At the end of the listing, the number of students and the class average is printed and punched.

### 2. FORTRAN Programs

a. Multiple Scores per Single Card. This short program reads the student number, name, and course identification in the usual format (Fig. 2) followed by a maximum of 14 three-digit fields for numeric scores in the contiguous cc 31 through 72. The 14 (or less) scores are added and the (crossfooted) total printed along with all information from each card on the same printline. At the end of the printout, the number of students, total scores, and average score is printed.

b. Single Score per Card. This program sums the scores from a series of cards for each student, printing all information from the cards in the usual format (Fig. 2), with the score punched into cc 73 thru 77, right-justified so that the last or units digit ends in cc 77. The program accepts the punched output from the grading programs described (in Section VI. B) after proper merging as described in Section VII. A.

A header card containing up to 54 characters of identification should precede the student cards. A blank card should follow each class (of cards) for which separate statistical data is desired. The end card after all decks should have a 9-punch in cc 80.

Summary score cards are punched for each student including the usual identification, number of scores, the total score per student and the student's average. At the end of a class of students, the total score for all students in the class, the number of students in the class, the class average, and the standard deviation are printed and punched for the record.

## VIII. APPLICATION

As this system has been developed over a period of several years, portions of it have been utilized for varying periods of time. The earliest to be used was the punch cards with student ID information as answer cards for large class quizzes. These were graded manually and returned the following class period. In this way, they served a dual purpose. Cards remaining upon issuance indicated absent students on the day of the quiz, and those remaining after grading indicated absent students on the following class day. This was a much more efficient method of taking a required roll call for 100 to 200 students in a large lecture hall without consuming class time. Since the original deck was in alphabetical order and the returned deck easily machine alphabetized, issuance of the cards was very rapid. Of course, they also served as the scorecard record.

Subsequently, mark sense cards were used in similar fashion for pop quizzes of the objective answer type, each card serving for several consecutive quizzes. Cards for absent students were gangpunched with incorrect entries; marks were converted into punches so the student could not change previously incorrect answers. The students were given the answers for each previous quiz and when the card was filled, it was graded and the total score printed for student verification. The cost of cards, including punching and printing, is estimated at about one cent each, excluding my labor.

Mark sense cards have been used for exam purposes by many instructors at our campus. I have used them for exam, quiz, homework, and lab unknown reporting for the past seven years, involving about 70 class sections and over 3000 different students. Most students did not appreciate the more frequent quizzing possible with an automated system, but did appreciate the more uniform and accurate grading, and the opportunity to compare their answers and performance with all of the members of the class from the posted computer listings. Some rebelled at using the Hollerith code printed on their cards for punching letters, but the more technically minded appreciated exposure to data processing principles, and learned how to feed cards to the IBM 1620 computer.

Port-a-punch cards were compared with mark sense cards during the last two years. Their advantage lies in eliminating the mark sense conversion process. However, they were more difficult to erase--an opaque tape was required. I prefer the Port-a-punch for answering homework, but mark sense cards for classroom testing.

Printing costs for "Custom Homework" and key are estimated at five cents per page on the 407. Costs could be more or less on a computer, depending largely on its printer speed. Grading costs are estimated at about a penny per card processed--again largely dependent on card reader, punch, and printer speeds.

The major advantages of the "Custom Homework" are that it eliminates copying, brings many more students in for needed assistance, and equitably rewards them much more (I gave half credit in some courses) for their efforts.

Additionally, all test questions were taken directly from the class homework, encouraging students to study other students' questions and practice other problems, and not just try to memorize computer printout answers to recognize from multiple-choice selections, as none appear on the tests.

I was particularly pleased with the initiative, interest, and effort the students put forth in their quest for the larger incentives that employment of this system allowed to be impartially and justly offered.

## ACKNOWLEDGMENTS

Implementation of this system was not a solitary operation. As a beginning programmer, I was greatly assisted by computer students Gale Capps, Don Erwin, and Dan Torkelson and data processing staff Perry Schmidt and Gary Austin. Computer Technology Chairman Dr. John Maniotes and computer center manager Don Kurtz were most helpful with their valuable suggestions and encouragements. The use of the equipment is especially appreciated, as is the partial financial assistance of the du Pont Small Grants Program of the Chemical Education Division of the American Chemical Society. The Journal of Chemical Education is thanked for publishing summaries during the development of the system [11, 12].

My heart goes out to my wife and daughters who have put up with my irregular hours, helped type, keypunch, wire boards, and kept the machines fed with cards, and, of course, to my long-suffering students who have been privileged to be patients of the system during its gestation.

The reader may reap the benefit of all these efforts by writing for copies of each source program, or a card deck of either a source or object program, enclosing postage (most decks are under one pound) and $1 handling charge. Please indicate the type of equipment (and its capacity) you have available for best suggestive hints for its use.

## APPENDIX

### Printing by the 407 Machine--Board Wiring Program

The wiring programmed here is an adaptation of that found on pp. 90 and 91 of IBM Manual A24-1011-2 [6]. The wiring will be discussed primarily by quarter sections of the board shown in Fig. 8, although some functions are wired from one quarter of the board to another.

Letters in CAPS designate names of areas found on the wiring board. Paragraph numbers refer to correspondingly numbered areas of Fig. 8.

#### 1. Upper Left Quarter

(1) At FIRST READING, card columns 10-12 are wired to COMPARING ENTRY. At SECOND READING, the same card columns are wired into their COMPARING ENTRY. Jack plug wires bridge the two entries at the middle area called COMPARING EXIT. Since these columns in the cards contain the question identification, an electrical impulse exists every time there is a difference perceived in the question identification.

(2) A wire from the left side of the exit delivers this impulse to the MInor PROGRAM START area to the right.

(3) Another wire from the right side of the exit delivers this impulse to the SKIP ConTRol hole (hub).

(4) The PROGRAM START O-FLOW hub is wired from O-FLOW N for activation of various steps after skipping to a new page of the printout.

(5) O-FLOW N is wired to channel 1 of CARRIAGE SKIPS to advance the new page to the first printline via the control tape.

(6) At FIRST READING, cc 14, which contains a - in the first header card designating information to be stored for page headings, is wired to INput STORAGE unit A, for such cards. IN STORAGE will be discussed further.

(7) The first four positions of UNIT A STORAGE ENTRY receive the heading information from cc 17-20 of the first header card through these four wires.

(8) This stored information reaches print positions 1, 2, 4, and 6 of TRANSFER PRINT ENTRY and also positions 103, 104, 106 and 108 for printing on both the left and right sides of the top of each page of the printout. Both of these are wired from UNIT A STORAGE EXIT positions 1-4 and are single spaced between the day & month and between the month & year positions in the printout.

(9) The NORMAL PRINT ENTRY positions 15-80 are each wired from SECOND READING card columns 15-80 for printing the text of each question of the homework.

(10) Card columns 1-12, which contain the answer and question identification, are wired into NORMAL PRINT ENTRY positions 91-112 to print this information on the right side of each page for cutting off as the instructors key for the homework.

(11) Note that card column 13 of SECOND READING is wired into the right-most position of COUNTER ENTRY number 1 (or 4A) in the lower left quarter of the panel. This enters into the question count the value of the answer length punched in cc 13 of the question card.

2. Lower Left Quarter

(1) ENTRY O and & - are wired into the common BUS bar for emitting impulses into the ZERO PRINT CONTROL in the lower right quarter of the wiring panel.

(2) COUNTER ENTRY 15 (3D) receives a 1-impulse as needed for page numbering.

(3) COUNTER EXIT 15 (3D) is wired to two sets of COUNTER CONTROLLED PRINT positions 8-10 and 110-112 for printing the page number simultaneously on both the left and right heading of each new page.

(4) The information entering counter 1 (4A) is carried from 1 COUNTER EXIT of 4A to two sets of print positions for printing on each side of the page simultaneously in COUNTER CONTROLLED PRINT positions 1-4 and 86-89 on the same printline as the question and gives the student the answer length required for that question. The cumulative total of answer positions prints on the next line after completion of a question.

3. Lower Right Quarter

(1) The impulses from the common BUS bar at the bottom of the Lower Left Quarter are received into the upper portion of ZERO PRINT CONTROL positions 1 & 2. Successive positions 3 & 4 are jackplugged in pairs throughout the 112 print positions used in this procedure. Note the use of the underlying aluminum foil (insulated from the frame of the panel) which carries the impulse between otherwise separate jackplugs. This wiring causes all separated and leading zeroes as well as + and - to print in any position. Otherwise, printing of these is normally suppressed by the 407 machine.

(2) COUNTER ENTRY 15 (3D) receives a 1-impulse for page numbering and incrementing as needed from CC (card cycles).

(3) The IN STORAGE hub of unit A is jackplugged to accept ALPHABETic information upon encountering a - punch from FIRST READING of cc 14 wired to the X hub.

(4) IMMEDIATE reading STORAGE OUT hub A is wired from the T-hub (transferred) of a relay in CO-SELECTOR 1 in the upper right quarter.

(5) STEP 1 PROGRAM MINOR is wired to cause counter 4A to READOUT after each minor step (change in question ID). This numbering of the question, sequentially based on the answer length, follows the last printline of each question.

(6) READOUT of counter 15 (3D) results when its wire carries an impulse from the T-hub of CO-SELECTOR 1 (see upper right quarter).

(7) O-FLOW 1 is wired to O-Flow END to terminate overflow operations after page headings, numbering, etc.

(8) OF CPL 1 (overflow couple 1) is wired to CO-SELECTOR PICKUP 1 [see (2) in upper right quarter)]. When channel 12 of the carriage tape is

encountered, this actuates the overflow, suspending NORMAL PRINT operations until all overflow operations are completed.

(9) ALL CYCLES impulses are split-wired to the Common hub of the relays of CO-SELECTOR 1 for use as needed and to the Common hub of Pilot SELECTOR 13 (see upper right quarter).

4. Upper Right Quarter

(1) SPACE 1 is wired to normally single line space the printing except at minor control (question) breaks.

(2) CO-SELECTOR PICKUP 1 is transferred [by wiring from the OF CPL 1, see (8) in lower right quarter] everytime the bottom of one page is sensed and a SKIP is made to the first print line of the next page.

(3) A relay of CO-SELECTOR 1, when in the transferred position, sends an ALL CYCLES impulse thru its T-hub to TRansfer Print stored information from unit A of storage.

(4) The T-hub of a relay of CO-SELECTOR 1, when transferred, allows the ALL CYCLES impulse to flow through it to the IMMEDIATE OUT OF STORAGE unit A.

(5) A CARD CYCLES impulses enters information from cc 13 of each card into the PLUS ENTRY of COUNTER 1, (4A) to increment the question number of the printout by the value of the answer length and to cause the printing of the answer length on both sides of the page of question text.

(6) Another relay of CO-SELECTOR 1, when transferred, transmits an ALL CYCLES impulse thru its T-hub to READOUT counter 15 (3D) for printing the incremented page number with the heading of each new page.

(7) The last relay of CO-SELECTOR 1, when transferred, transmits an ALL CYCLES impulse to flow from the Common through the T-hub to PLUS ENTRY of COUNTER 15 (3D). This increments the page counter after each overflow time, i.e., after printing of the page number at the top of the new page.

## REFERENCES

1. IBM Reference Manual A24-0520-3, 1965.
2. IBM Accounting 10,000 Division Code for Proper Names X21-5114, reprinted from Practical Applications of the Punched Card Method

in Colleges and Universities, (G. W. Baehne, ed.), Columbia University Press, 1935.

3. IBM, Reference Manual A24-1005-2, 1960.
4. IBM Reference Manual 224-6384-2, 1958.
5. IBM Reference Manual A24-1002-2, 1959.
6. IBM Reference Manual A24-1011-2, 1965, pp. 13-14, 15-16, 66-68, 90-91, 117-120, 150-153.
7. IBM File No. S360 (Mod 20) - 28, Form C26-3600-5, 1967.
8. J. Maniotes, H. B. Higley, and J. N. Haag, Beginning FORTRAN, Hayden, New York, 1971, pp. 169-171.
9. N. A. Lange, Handbook of Chemistry, Handbook Publishers, Sandusky, Ohio, 1967, pp. 1436-1450.
10. R. C. Weast, Handbook of Chemistry and Physics, The Chemical Rubber Co., Cleveland, Ohio, 1970, pp. D-167-9.
11. N. Doyal Yaney, J. Chem. Ed., 44, 677 (1967).
12. N. Doyal Yaney, J. Chem. Ed., 48, 276-7, (1971).

# AUTHOR INDEX

Numbers in parentheses are reference numbers and indicate that an author's work is referred to although his name is not cited in the text. Underlined numbers give the pages on which the complete references are listed.

A



B



C

D



E



F



G



H



I

J



K



L



M



N



O



P

## SUBJECT INDEX

Q



R



S



T